Inside the Designer

Understanding imagining in spatial design

Marisha McAuliffe PhD

Marisha McAuliffe

Inside the Designer: Understanding imagining in the spatial design

National Library of Australia Cataloguing-in-Publication Entry

Creator: McAuliffe, Marisha, author.
Title: Inside the designer : understanding imagining in spatial design / Marisha McAuliffe.
ISBN: 9781329778566 (paperback)
Subjects: Space (Architecture); Architectural design; Interior decoration; Design—Research; Creation (Literary, artistic, etc.)
Dewey Number: 720.1

First Published in 2016
by Oxford Global Press
London (UK)
Sydney (Australia)
www.primrosehall.com

 **These works have been blind peer reviewed.**

CHAPTER 1: TOWARDS IMAGINATION IN DESIGN 7
An Outline of the Book's Chapters 11
CHAPTER 2: EXPLORING DESIGN METHODOLOGICAL RESEARCH 13
Introduction 13
Design Process 14
Design Thinking and Cognition 21
Imagining 25
Creative mental synthesis 38
Summary and discussion 40
CHAPTER 3: THE PRESENCE RESEARCH FRAMEWORK 43
Introduction 43
Presence Research Framework 44
Technology Mediated Presence 50
Non-mediated (physical) presence and (non-technology) mediated presence52
The Presence Divide 69
Conceptual Presence 81
The mental model 81
Suspension of disbelief 83
Facilitating mental simulation 87
Discussion and Conclusion 89
CHAPTER 4: INTO THE STUDY: A METHODOLOGY 92
Introduction 92
Grounded Theory as a Methodological Framework 92
Summary 101
CHAPTER 5: INTO THE STUDY: EMPIRICALLY GROUNDED IMAGINING 102
Introduction 102
Imagining and Designing 102

Empirically Grounded Categories of Imagining 110

Visual Imagining 110

Aesthetic Imagining 118

(Con)textual Imagining 121

Factors Influencing Imagining 123

Conclusion 126

CHAPTER 6: INTO THE STUDY: THE SPATIAL DESIGN IMAGINING (SDI) MODEL 127

Introduction 127

Imagining and design methodology research 127

Imagining and presence research 134

The Spatial Design Imagining model 137

What is imagining? 140

CHAPTER 7: CONCLUDING THE STUDY: A DISCUSSION AND CONCLUSION 144

Introduction 144

Significance and contribution in enhancing knowledge of imagining and its role in spatial design 145

Implications for spatial design education specifically design studio teaching 147

Wider implications and contribution 154

Conclusion 155

REFERENCES 158

Chapter 1: Towards

Imagination in Design

> *Of all the questions we can ask about design, the matter of what goes on inside the designer's head is by far the most difficult and yet the most interesting and vital (Lawson, 1980, p. 94).*

More than three decades after Lawson (1980) made this statement, the matter of 'what goes on in a designer's head', or imagining and mental problem solving, remains just as mysterious and just as pertinent, possibly more so given the social and environmental challenges facing humankind. It is within this context that I introduce this book.

This book is concerned with expanding understanding of imagining in the spatial design disciplines of architecture and interior design. Its genesis is in doctoral studies that I undertook in 2010. With this knowledge you'll come to appreciate the rather formal nature of this book and the structures that make it a research based study.

The impetus for this book began in the interior design studio with a desire to better understand how I could enhance the learning of my design students. Informing this were observations and concerns about the initial stages of designing and how much was developed 'in the student's mind' (the imagining process) before it was externalised and made available for evaluation and comment. Despite attempts to help my students develop drawing skills quickly so as to represent and expose their thinking and imagining (which is what we are led to believe we need to do as design educators), there was an inherent reluctance, even resistance, by the students. While this may be due to several factors including lack of confidence and inability to represent ideas, interestingly, further reading showed this to be a preference for some well-known designers such as architect Frank Lloyd Wright who proposed:

> Conceive the building in the imagination, not on paper but the mind, thoroughly – before touching paper...Let it live there – gradually taking more definite form before committing it to the drafting board (Wright, 1928).

Reading this then prompted questions about what we needed to focus on and how we might in our educational role facilitate conceptualisation in the imagination in design education. As pointed out by Athavankar (1997), this was a question to which at that time there were very few, if any, answers:

> [Design] education has neglected the development of visualization and imaging abilities, not fully realizing their potentials as well as implications for creative pursuits. There are no conscious attempts to promote the development of imagery and abilities to control images voluntarily and thereby facilitate problem solving (Athavankar, 1997, p. 39).

To better understand the context of this, an in-depth review of design methodological literature was undertaken which is presented in Chapter 2 of this book. As outlined, the review pointed to early research concerned with developing prescriptive models of design in order to improve process efficiency and product performance. Underlying this was an understanding of design as a mechanistic, sequential activity. When these new models failed to achieve the desired outcomes, especially in the spatial design areas such as architecture and interior design, attention turned to better understanding the nature of the design task, and with this a new conception of designing as a heuristic, satisficing activity emerged. As the review shows, this was facilitated through research concerned with the nature of design problems and design as a (creative) cognitive activity. Design problems were identified as being ill-defined and 'wicked' due to their complexity and future oriented nature, demanding a generative way of reasoning involving abductive, as well as inductive and deductive, thinking. While there was some associated research to do with visualisation and modelling, this was minimal, prompting Kokotovich's (2000) call for more systematic methodology in order to extend the views expressed in the design literature relating to creative mental synthesis. He argued that whilst there has been extensive research over the past thirty years in perceptual psychology, the research had not specifically addressed issues in design thinking. He proposed that "[D]eveloping a detailed understanding of creative mental synthesis. This book will serve to support design education, and therefore the improvement of design practice. Cognitive processes are central to the process and practice of design. Consequently, it is important that some of these cognitive processes be identified and understood" (Kokotovich, 2000, p. 2).

In response, there has been some recent research focussing on abductive thinking and the notion of framing (for example, Paton & Dorst, 2011) and imagining as a process (for example, Folkmann, 2010) but as emphasised in the book, this focuses almost exclusively on thinking and problem solving processes in the industrial design discipline, downplaying the experiential and aesthetic dimensions to designing and the role of imagining in facilitating embodied creative mental synthesis, especially in spatial design.

To further understand all this, this work is focused through a key research question and hence takes the shape of a structured study. It is important to appreciate at this point that this book had its genesis in a formal doctoral program that was completed in 2013.

What is imagining in the spatial design process?

Specifically, this research question requires me to ask and thus investigate:

- What is imagining as experienced by interior design and architectural practitioners?
- What extant theory is relevant and, when considered with current empirical research, is likely to expand current knowledge of imagining?
- What are some initial implications for spatial design education and future research?

In responding to these questions, the research approach for this book was informed substantially by Grounded Theory methodology. Initial research concerned with finding an appropriate methodology revealed how Grounded Theory is particularly useful for exploring an issue faced by a specific community or group (Glaser, 1992), especially a qualitative issue situated as an integrated problem requiring an explorative, abductive-inductive approach for the generation of new insights. Philosophically, then, the research is situated in the interpretive paradigm grounded more specifically in a constructivist context concerned with how the complexities of the socio-cultural world are experienced, interpreted and understood at a particular point in time (Mills, Bonner & Francis, 2006a). Subsequently, what underpins this research is a strong congruence between a personal philosophical position, the stated aims of the research, and the methodological approach.

As explained in detail in Chapter 4, the research process developed iteratively over several years. Initially, the study commenced with a focus on student designing and their experience of designing in the early stages of the process. When very little emerged from the data analysis from the first exploratory stage, the study turned to postgraduate students. While this produced more data and understanding about the experience of designing, again it was limited and severely constrained by the students' little or no practice experience. At this point, the decision was made to involve experienced practicing architects and interior designers. During this stage data were collected via questionnaire from 54 practicing designers and 10 follow up interviews. In parallel with this process, data were also collected from design methodological literature and presence literature at various points across these stages. Unlike traditional research where the literature review occurs chiefly at the beginning to establish a context, in this study, literature and existing theory were accessed when directed to as a result of previous analysis of literature and empirical data. In accordance with Grounded Theory methodology, the process was informed by the principles of theoretical sampling and constant comparative analysis to clarify and refine emerging themes until they were 'saturated'. As referenced by Birks and Mills (2011):

> In Grounded Theory research, the aim is to build theory through the construction of categories directly from the data. Through 'theory-directed' sampling, you are able to examine concepts from various angles and question their meaning for your developing theory (Strauss, 1987, p. 276 in Birks & Mills, 2011, p. 69).

An integral aspect of theoretical sampling for this study was the inclusion of extant theory from presence research. In this respect, it played a major role in extending an understanding of how one engages with complexity through immersion and how this can be mediated through technology or otherwise as will be described more fully in Chapters 3 and 7. Of particular note is how 'less artificial' mediation and a lack of detail and stimuli appear to support richer aesthetic engagement. Empirical data collected from the designers also played a significant role and were central in the development of the taxonomy of categories of imagining, a major outcome of the study.

Having now made these introductory statements, I now provide an outline of the remaining chapters.

An Outline of the Book's Chapters

The book has seven chapters. The first, **Chapter 1**, as presented here, describes the impetus for the study. This is followed by an outline of the research questions, underpinning philosophical position and research approach. In addition, it gives a brief overview of the major outcomes and their significance when explored as implications for spatial design education and their potential to contribute to the discipline's body of knowledge. **Chapter 2** presents an overview of design methodology research as a context for a closer examination of research focussing on design thinking, cognition and concepts that connect with the phenomenon of imagining. **Chapter 3** then provides an overview of presence research including its history and areas of investigation that are of particular relevance to imagining such as the role of cognition in presence research. In **Chapter 4** a description of the methodology, research process and research plan are provided together with justification of the decision to use Grounded Theory as the main methodological framework. Issues of research quality and rigour are also addressed. **Chapter 5** is concerned with the outcome of the analysis of empirical data obtained from the designer participants of the study. The first section describes how designers regard imagining in the design process. The second section presents the outcome of a closer examination of imagining and the emergence of empirically grounded categories of imagining. **Chapter 6** focuses on the main outcome of the study – the Spatial Design Imagining (SDI) Model. It outlines the development of the model according to the integration of extant theory from design methodology, the empirically grounded categories of imagining and the extant theory from presence research. **Chapter 7** concludes the book highlighting the significance and contribution of the study in enhancing knowledge of imagining and its role in spatial designing. It provides further demonstration of its potential applicability by exploring the implications of the findings for spatial design education, in particular design studio teaching and learning. It then suggests some areas of wider relevance and contribution and concludes with a number of recommendations for implementation and future research.

Chapter 2: Exploring Design Methodological Research

Introduction

This chapter presents an overview of design methodology research. It commences with an historical overview of research concerned with the general area of design process. While the main focus of this book is imagining in the spatial design disciplines of architecture and interior design, the review cannot ignore early research in engineering and industrial design and the way this has influenced and dominated methodological research in architecture and interior design. Further to this, an examination of design process research also provides a context for a review of research focussing on design thinking and cognition, which, in turn, provides a platform for a closer examination of imagining. These aspects of designing are presented in Figure 2.1 embedded in the broader area of design practice. While there is a body of literature concerned with architectural and interior practice, the focus of this review is chiefly on the process undertaken by the designer during the initial concept and design development phases of a design project; it is during these stages that imagining assumes a significant role. It is acknowledged that while communicating and representing "what goes on in the designer's head" is an important part of the design process, design communication and representation is beyond the scope of this literature review. Additionally, the review of educational theories surrounding learning (in general), design education and creativity are also considered to be beyond the scope of this book.

This chapter, together with the following chapter which includes a review of presence research, highlights the extant theory considered in relation to the findings of the empirical study described in Chapter 5 and which contributed to the generation of the Spatial Design Imagining (SDI) Model presented as the outcome of this book in Chapter 6.

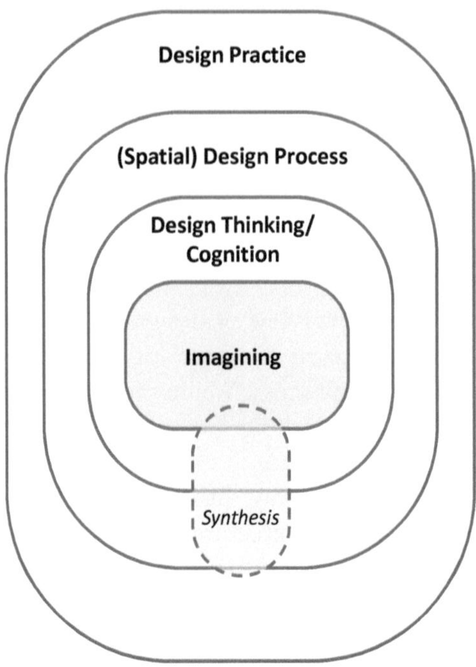

Figure 2.1: Literature review context for design process, thinking and imagining

Design Process

For the purposes of this research, design process is described as *the work undertaken by a designer or design team in generating a proposal related to a building or interior environment.* In design practice, this proposal is further documented and developed contractually to effect its realisation for use by the client.

Several authors including provide overviews of methodological research in design and architecture up to the late 1990s, which reveal an initial emphasis on interventionist and prescriptive models to inform more effective and efficient designing and enhanced product performance. When these models failed to have the anticipated outcomes, researchers turned their attention to better understanding the ill-defined nature of the design task and how designers cognitively manage ill-definition as part of the design process.

Design as a mechanistic, prescriptive activity

Following developments in mathematics, logical methods and behaviourist psychology in the early 1900s, methodological research in design gained momentum with the first conference on design methodology in London in 1962 attended by a range of disciplines including engineering, industrial design, and architecture. Despite the presence of architecture, some of the first models developed were by engineers such as Asimow (1962), and the industrial designers J. Jones (1963) and Archer (1965). Common to these 'first generation' (Rittel, 1972), 'glass box' (Jones, J. 1970) analysis-synthesis-evaluation models was a conception of design as a rational process involving a systematic decomposition of the problem into sub-problems. "Design methods people were looking at rational methods of incorporating scientific techniques and knowledge into the design process to make rational decisions" (Bayazit, 2004, p. 18). Similarly for Archer (1980), "design research is a systematic process whose goal is knowledge of, or in, the embodiment of configuration, composition, structure, purpose, value, and meaning in man-made things and systems" (Archer, 1980, pp. 30-47). At that time, logical process and scientific principle were integral concepts in design methodology research. For Fielden (1963), design was understood as "the use of scientific principles, technical information and imagination in the definition of a structure, machine or system to perform pre-specified functions with the maximum economy and efficiency" (Fielden, 1963, p. 60).

To help with combining sub-problems simultaneously rather than sequentially into an overall optimum solution, Luckman (1969) drew on the work of Archer (1965) to develop the procedure known as 'the analysis of interconnected decision areas' (AIDA). In another vein, Thornley (1963) focussed on systematising the initial stages of designing from a practice perspective and is attributed with what we now know as brief development, concept development, schematic development, and design development.

In 1963, researchers began utilising computer technology to produce computer graphics systems in wireframe and polygonal modelling. Moseley (1963) was one of the first researchers who developed and utilised computer technology to design the first layout optimisation program for a hospital operating unit. Researchers such as Simon (1969) and Newell and Simon (1972) increasingly regarded computer science as a way of managing complexity and the inability of humans to cognitively deal with complex situations. Mitchell (1977, 1990) drew extensively on artificial intelligence, cognitive science and the theory of computation in attempts to demonstrate how the structure of architectural design reasoning could be understood logically. In the main, however, such attempts were not fruitful with Radford and Woodbury (1991) delimiting the use of computation in design to the formulation of mathematical systems for design representation and generation. Even with recent advances in computer science and Building Information Modelling (BIM), the use of computational tools in the spatial domains of architecture and interior design largely remains restricted to that outlined by Radford and Woodbury in 1991.

Limitations with the conception of design as mechanistic and prescriptive began to appear in the 1960s in line with increasing awareness of the ill-defined nature of the design task, what this means in terms of the outcome and how this invokes a particular way of thinking and doing.

Design as a heuristic, satisficing activity

Studies by Darke (1979), Lawson (1979), Zeisel (1981, 1984), and Schön (1983) appear to be among the first to differentiate the architectural design process from other activities. In relation to research by Darke (1979), her approach to managing ill-definition was to apply what she terms 'primary generators' which are concepts about the problem which help constrain the problem space to be searched. The constraints are then used to test possible 'solutions' or 'conjectures'. While Darke (1979/1984) uses the term 'primary generator', it has also been referred to as the 'kernel idea', 'key concept', 'early solution conjecture', 'primary position' and 'guiding theme' (Cross, 2001a, 2004b; Guindon, Krasner & Curtis, 1987; Kant, 1985; Lawson, 1980; Rowe, 1987; Ullman, Dietterich & Staufer, 1988). Subsequently, the ensuing process has been qualified as 'position-driven' design, 'early fixation', 'premature commitment', 'early crystallisation', or 'solution fixation' (Ball, Evans & Dennis, 1994; Cross, 2001a; Goel, 1995).

In this respect, then, researchers moved from an understanding of designing as a step-by-step sequential process to an understanding of it as fluid and dynamic. They began to realise that designers constantly generate new task goals and redefine task constraints; and even if they are cognisant of prescriptive models, designers do not follow them systematically (Akin, 1979/1984; Carroll & Rosson, 1985; Cross, 1984; Dasgupta, 1989; Visser, 1987a).

Extending on the work by Darke (1979) and Lawson (1980), Zeisel (1981, 1984) proposes that there are three elementary design activities: imaging, presenting and testing, and that these are undertaken iteratively (Figure 2.2). Here, potential ideas are presented, most commonly, externally through drawings of varying abstraction in order to be 'tested' by the designer or used to invite feedback by others. Adding to this, Schön (1983) asserts that designers formulate and test rules in the form of visual communication: drawings and models. "Rules in design are largely implicit, overlapping, diverse, variously applied, contextually dependent, and subject to exceptions and critical modification" (Schön, 1983, p. 133). For Schön (1983), situations of uncertainty, novelty and complexity, whether in design or other areas, demand a process of reflection in and on action; in other words, a knowing-in-action.

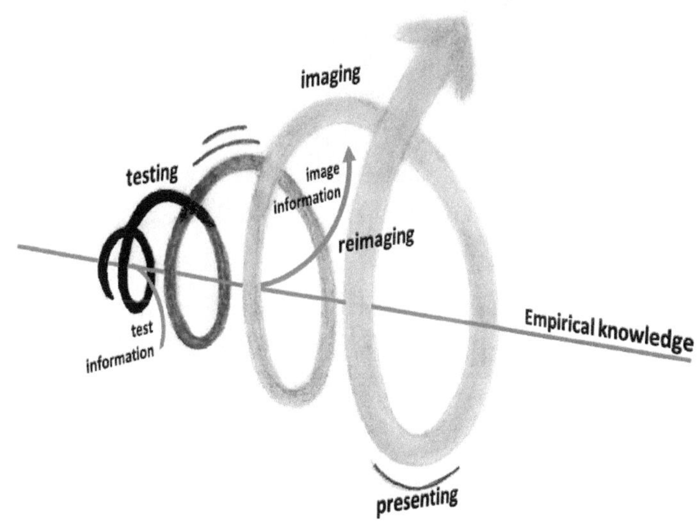

Figure 2.2: The three elementary designing activities based on Zeisel (1984)

For Cross (1984), this is described as a 'designerly way of knowing': "there are designerly ways of knowing distinct from the more usually recognized scientific and scholarly ways of knowing" which involve tackling ill-defined problems as well as well-defined problems, a constructive (or abductive/productive) mode of reasoning; and the use of codes for communication and translation from the abstract to the concrete (Cross, 1984, p. 226). While this notion of abductive thinking will be examined more fully in the subsequent section, the following discussion is useful in providing an initial description differentiating it from inductive and deductive forms of reasoning: "Abduction, or as we have it production, is the only logical operation which introduces new ideas; for induction does nothing but determine a value; and deduction merely evolved the necessary consequences of a pure hypothesis. Thus production creates; deduction predicts; induction evaluates" (March, 1984, in Teeravarunyou & Teixeira, 2002, p. 122).

Summarising designerly ways of knowing in later work, Cross (2001a) describes how working in novel situations with incomplete information demands that designers use imagination and constructive forethought in a heuristic way aided by drawings and modelling. To this end, his understanding was most likely informed by other research conducted in the 1980s and early to mid-1990s. The 1980s saw the development of several models of architectural designing, including a descriptive model by Goldschmidt (1983), providing the basis for future research in the 1990s and a greater focus on conjecture, and prescriptive models by Heath (1984) and Akin (1986a).

More recently, Nelson and Stolterman (2003) take a holistic view characterising design inquiry as not compatible with the existing domains of inquiry. They frame the practice of design as a broad culture of inquiry and action, asserting that rather than focussing on problem solving to avoid undesirable states, designers work to frame problems in terms of intentional actions that lead to a desirable and appropriate state of reality. In their view, design is viewed as a unique way to look at the human condition, and is understood through reflective practice, intellectual apperception and intentional choice. Their proposition is that a new philosophical culture in design not only applies to those fields focused on physical design and traditionally thought of when as design (for example, architecture, interior design, graphic design, software design), but encompasses other design areas including 'educational systems design'.

Essentially, they argue that this new culture promotes a 'design way of thinking', perceiving that design is the ability to imagine "that-which-does not-yet-exist" and make it appear in a concrete form as purposeful new addition to the real world. Taking this notion, they describe the need for imagination and judgment, where imagination allows us to create the "not-yet-existing" through a process of composing parts, functions, structures, processes, and forms in way that fits the design situation. Judgment is used to evaluate that composition to determine how well it fits (Nelson & Stolterman, 2003). However, enterprise also requires additional skills in design communication (both internal and interpersonal) to bring the composition into a perceivable form that allows judgment to occur.

According to Nelson and Stolterman (2003), there are four conceptual foundations of design: *the real; service; systems thinking;* and the *whole*, where 'the real' is intended to clarify what should be the express focus of design inquiry. The 'real' world, (while including the 'true' or natural world) is an artificial world, a created design. The design process concerns moving from general and universal to a specific design using a multifaceted form of inquiry (design inquiry), composed of the real, the true, and the ideal. 'Service' is the crucial element of design that distinguishes it from other traditions of inquiry. 'Systems thinking' is 'the' organising element in design reasoning, providing a different perspective to traditional design theory and practice. Finally, design is to be understood as a complete 'whole', not as a series of tasks, or steps.

They propose that the fundamentals of design include the essential skills of design inquiry and practice.

These essential skills are: desiderata, interpretation and measurement, imagination and communication, judgment, composition, and production and care taking. Desiderata (those things that are desired) is the process of giving direction and the importance of intention as an initiator of design action.

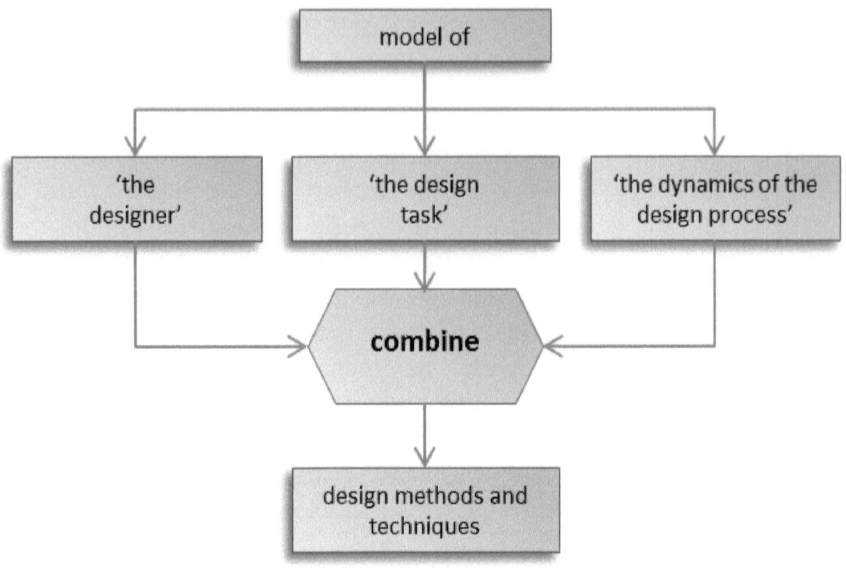

Figure 2.3: The three dimensions of design process based on Roozenburg and Cross (1991)

Interpretation and measurement concerns appreciative judgment about what is to be considered and the exploration of possibilities, leading to a compositional interpretation. Imagination and communication concerns design creativity, and describe the phases of the design communication process. Judgment defines various types of judgments (appreciative, quality, instrumental, and framing) in the design process and delineates the differences between judgments made by the client and those made by the designer. Composition is the phases in the design process from the 'parti' (the 'ah-ha' experience) through to the design innovation or 'ultimate particular'. Production and care taking is a focus on the care of the material of the design and management of the production process.

As indicated previously, the research summarised in this section and methods or models developed acknowledge a specific type of dynamic in relation to the designer and the design task. Roozenburg and Cross (1991) convey this graphically in Figure 2.3. According to Dorst (2004), any method for aiding design activities contains assumptions about all three dimensions.

The realisation of design as a heuristic, satisficing activity that emerged in the 1960s persists to this very day. As Kolko (2010) observes, "Designers, as well as those who research and describe the process of design, continually describe design as a way of organizing complexity or finding clarity in chaos" (Kolko, 2010, p. 15). How designers do this has been the subject of research focussing more specifically on the designer, and their design thinking and cognition in relation to the design task at hand.

Design Thinking and Cognition

"Of all the questions we can ask about design, the matter of what goes on inside the designer's head is by far the most difficult and yet the most interesting and vital. This leads inevitably into the realm of cognitive psychology, the study of problem solving and creativity, in short thought itself" (Lawson, 1980, p. 94).

As the following discussion will highlight, interest in more fully understanding design thinking and cognition followed the earlier discussions in designing; thus increasing the realisation of the ill-definition and uncertainty managed by designers when designing.

The nature of design problems

The understanding of design as that involving a special kind of problem solving is largely attributed to foundational research by Reitman (1964) and Eastman (1970), followed by studies by Rittel and Webber (1973) and Simon (1973). Eastman's work, for example, points to a process in problem solving involving matching appropriate forms of representation such as words, diagrams and plans with particular information processing activities (Eastman, 1970, p. 30). This research and that of Goldschmidt (1983) and Mackinder and Marvin (1982) suggest a reliance of past personal experience for generating constraints for problems possessing no defined criterion for testing a proposed solution. Such problems are described by Rittel and Webber (1973) as wicked and incorrigible, and by Simon (1973) as ill-structured. With no formulation for establishing what is or what is not optimum, designers wanting to change existing situations into preferred ones (Simon, 1981) can only work towards what Simon (1969) refers to as a 'satisficing' outcome.

Following on from this research, the ill-defined nature of design problems continues to be acknowledged (Akin, 2001; Michalek & Papalambros, 2002; Ormerod, 2005), as does the understanding in architecture that the process can only be at its best satisficing (Buckingham & Shum, 1997; Conklin, 2006). As Akin (2001) states, rather than this occurring in every discipline, designers from different disciplines vary on this point: while architects indeed proceed to satisficing, engineering designers tend to adopt more objective methods in their selection among possibilities and may proceed to optimisation. Further to this, rather than one solution, which would be 'the' correct solution, design problems have several acceptable solutions, which are more or less satisfying.

According to Wade (1977), designers work towards a satisficing outcome by drawing on existing exemplars and/or by applying constraints to the point where a certain outcome appears feasible. Coupled with generic knowledge (including from design methodology, the application domain, and the technical domains that underlie the design project) reuse of knowledge (from specific previous design projects) through analogical reasoning has been observed in many cognitive design studies as a central approach in design (Ball & Christensen, 2007; Ball, Ormerod & Morley, 2004; Bhatta & Goel, 1997; Visser, 1995, 1996). While the majority of examples of reuse concern software design (Visser, 1987b), this technique is also associated with product design (Visser, 1995) and architecture (Demian & Fruchter, 2005).

Lawson (1980), one of the most quoted researchers within the design theory area, argues that designers rely on episodic memory rather than general principles or problem-solving algorithms (that is, process models) for generating design 'solutions'. Episodic memory includes direct experience with designed artefacts (designing them or experiencing the designs of others) and vicarious experience with descriptions and representations of designed artefacts. More recently, Lawson (2004) explains how "precedent stored in the form of episodic memory", either of one's own design or of the designed artefacts of others, "is used by experts to recognize design situations for which gambits are available", where gambits are defined as "patterns known to have certain properties and to offer certain capabilities" (Lawson, 2004, p. 443).

Design as a (creative) cognitive activity

Not only are design problems 'wicked' because of their substantive complexity, they are 'wicked' and ill-defined because they are associated with producing something for the future. As described previously, design involves conjecture; it is productive and generative, in other words, it is a creative process inviting closer examination of the cognitive processes underlying the individual's capacity to generate new ideas or understandings.

Many theories in design research are dualistic knowledge models to explain design knowledge frameworks: the rational, problem-solving paradigm of design based on a techno-positivist framework developed by Simon (1969) and the alternative constructivist epistemology, proposed by Schön (1983), based on human perception and 'reflection-in-action'. However, the limitations of both models have been recognised and challenged by several authors, the most significant being Snodgrass and Coyne (1992, 2006). The polarity between the rational, systematic approach of design science and that of intuitive, tacit, practice-led approaches, described by Coyne and Snodgrass as 'design's dual knowledge book', has underpinned much of the discourse about design research over the last forty years.

Taking inspiration from Schön's (1993) musings on metaphor, Snodgrass and Coyne (2006) argue that we need to become aware of how the metaphors we use often determine how we think about design, suggesting that metaphors as an alternative to models have a better fit and are better able to capture complex, situated and to a large extent embodied human experience, for example, of designing.

Their suggestion is that the "hermeneutic circle" is a better metaphor for designing than the dominant metaphor of problem solving because it does not "destroy the complexity, subtlety, and uniqueness of the design situation; or privilege or preclude aspects of the process, but rather respects their interdependence and interaction" (Snodgrass & Coyne, 1992, p. 72). Further, the process of interpretation does not necessarily stop at the first access to meaning, but rather "it can prompt a revision of the projection allowing other meanings to appear. Metaphors and models do not have a static, on-off meaning but are potentially capable of revealing multiple meanings that can be progressively disclosed by the back and forth meaning of the hermeneutical circle' (Snodgrass & Coyne, 1992, p. 68).

They perceive that design is a hermeneutical activity, and that interpretation is the core of architectural production and hence, architectural understanding. In proposing hermeneutics, this attempts to address the question of how understanding (essentially a cognitive process) occurs. This understanding involves the act of projecting self or understanding onto the activity being undertaken. Snodgrass and Coyne's (2006) hermeneutical circle is analogous to Schön's concept of the "reflective practitioner" who undertakes "reflection-in-action"; the relationship between thought and action. As Coyne and Snodgrass note: "Even a cursory examination of the protocol studies of Donald Schön indicates that the design process he describes works according to the dynamics of the hermeneutic circle, proceeding by way of a dialogic exchange with the design situation" (Snodgrass & Coyne, 1997, p. 22).

To design is to engage in a process of understanding through the generation and interpreting of design responses. This is essentially a process of interpretation and underlying every act of understanding is an "anticipatory projection of meaning" (Snodgrass & Coyne, 2006, p. 38).

Of relevance is the research on abductive as opposed to inductive and deductive thinking. As March (1984) observes, the model of explanation in natural science primarily utilises deductive inference, yet social science uses inductive inference. He writes, "Abduction, or as we have it production, is the only logical operation which introduces new ideas; for induction does nothing but determine a value; and deduction merely evolved the necessary consequences of a pure hypothesis. Thus production creates; deduction predicts; induction evaluates" (March, 1984, in Teeravarunyou & Teixeira, 2002 p. 122). As further explanation, where deduction shows that the premise provides a *guarantee* of the truth of the conclusion, and a conclusion that is so strong that, if the premise is true, it would be *impossible* for the conclusion to be false; induction is an argument in which the premise is supposed to support the conclusion in such a way that if the premise is true, it is improbable that the conclusion would be false. On the other hand, abduction could be thought of as the "step of adopting a hypothesis as being suggested by the facts...a form of inference" (Peirce, 1913, in Kolko, 2010, p. 19). In other words, "abduction can be thought of as the argument to the best explanation... It is the hypothesis that makes the most sense given observed phenomenon or data and based on prior experience" (Kolko, 2010, p. 20).

As such, in the abductive reasoning process in design, the various constraints of the problem begin to act as a 'logical' basis, and the designer's work and life experiences begin to 'shape' the abduction. The abductive process acts as inference or intuition, and is directly aided and assisted by personal experience.

Imagining

This section provides an overview of research dealing with several aspects related to the concept of imagining. These include: imaging, visualisation and mental imaging; mental models, simulation and imaging; and creative mental synthesis. Effort has been made to distinguish between these aspects despite the tendency in research for specific terms to be used interchangeably. As the review will reveal, there is very little research that focuses explicitly on imagining.

Imaging, visualisation and mental imaging

Previous discussion has included reference to Zeisel's model of designing and three fundamental activities involving imaging, presenting and testing. In this model, "Imaging means forming a general, sometimes fuzzy, mental picture of part of the world. In design, as well as in other types of endeavours, images are often visual; they provide designers a larger framework within which to fit specific pieces of a problem as they are resolved" (Zeisel, 1981, p. 7). Also described as or in association with is the term 'visual thinking' which Roam (2008) describes as "taking advantage of our innate ability to see—both with our eyes and with our mind's eye— in order to discover ideas that are otherwise invisible, develop those ideas quickly and intuitively, and then share those ideas with other people in a way that they simply *get*" (Roam, 2008, p. 4). Seeing an image in one's mind is also described as mental imaging.

References to mental imaging appear in the 1960s with Arnheim (1969) pointing out its significance in visual arts. Suggesting that there is a presence of abstracted visual information in memory and in making this point, he provides an example discussing how an artist can draw something from memory, going through a process of drawing and then checking that drawing against the mental image.

Essentially the artist, when asked to draw something from memory, would begin a drawing and then continually return to the drawing to make corrections as the abstracted visual image brought more and more information to the forefront of memory. Furthermore, Arnheim (1969) states that mental images function as one of three types of images: pictures, signs or symbols. However, sometimes one mental image may serve simultaneously in all three types of images, but essentially, all three types of images may actually act as a symbol for an idea or image that is far more complex than can be pictured. The picture image will not be an exact replica of the thing itself, states Arnheim (1969), but rather will have ambiguities causing the thinker to make value judgements in the absence of needed information.

Mental images acting as pictures have a low level of abstraction and may inform on such levels as shape, form, colour or movement (Arnheim, 1969; Damasio, 1999). A sign image portrays a particular concept or idea without actually showing the physical detail, but may suggest simply the qualities of that thing for which it is a sign (Arnheim, 1969). It is an abstraction only inasmuch as the sign itself is an abstraction of a more complicated idea or concept; letters, numbers and mathematical signs fall into this category of mental image. For example, a letter can be imagined as a sign, but the thought of the letter "A" may specifically be referring to a letter, a word, a grade or an evaluation.

Finally, in terms of mental images acting as symbols, Arnheim (1969) explains that a symbol is a further abstraction of a picture and stands for something much more complex than that which the symbol actually is. A good example of symbolism is that of music notes where a music note on a stave can communicate at a glance the pitch of the tone and the duration of the tone.

McKim (1972), in discussing design issues, views mental imaging as a component of visual thinking and stresses the need for learning to voluntarily control the transformations and manipulations of images in the design process. Studies undertaken by Sommer (1978) of architects' design process revealed the flexibility and non-material character of images and their ability to allow unusual transformations. Singh (1999) also undertook observational analysis of the architectural design process, identifying that mental imaging plays an important role in the design process and that first, the "flexibility and speed available in mental imaging is far more superior than sketching or modelling may offer enabling a student to experiment and choose between options at higher rate than sketching or modelling" (Singh, 1999, p. 4). Other indirect references to mental imaging can be found in literature to do with artificial intelligence (AI) and expert systems developed for design problem solving (Oxman, 1994).

Connecting more directly to conjecture, mental imaging is also understood to occur when a designer imagines or visualises the possible outcomes or solutions to a design problem using internal pictures (Block, 1981; Damasio & Damasio, 1996). Here imagery is a cognitive function of the brain, allowing the designer to 'see' and manipulate ideas and possible solutions (Block, 1981). To date, there still remains a debate among researchers as to whether mental imagery is pictorial or descriptive (Block, 1981), with Fodor (1975) and Kosslyn (1978a, 1978b, 1980) proposing that mental imagery involves visualising using pictures and descriptive terms. As there are no physical pictures in the brain, their argument is based on the idea that mental images represent in much the same way that a picture does. On the other hand, Pylyshyn (2001, 2002a, 2002b, 2003a, 2003b, 2003c, 2004) and Dennett (1969, 1978, 1981, 1991, 2002) argue that visualisation is strictly descriptive in nature with mental imagery sitting firmly as a linguistic representation.

In contrast, Damasio (1999) asserts that thought is made up largely of images. Whenever an individual speaks, writes a sentence, or draws an image, the thought begins with a visual image: "If those thoughts were not imageable, we would not know them and would not be able to manipulate them consciously (Damasio, 1999, p. 107). Indeed, images are the main content of thought. In discussing visual thinking, Damasio (1994) suggests that images of things that were experienced visually are more concrete than images of things made up. If the thing that is known is something that is experienced externally, the brain will record in detail all of the facts of what is seen. If the thing that is known is experienced internally it will be more vague or abstract, but these "are images nonetheless" (Damasio, 1994, p. 108).

Hoffman (1998), on the same topic of visualising and imagery, presents a compelling argument that all experiences are in part a construction of the mind. Following several studies of patients affected by brain damage, Hoffman developed a series of rules that the mind applies to constructing an understanding of visual input. Where there has been damage to certain parts of the brain, a patients' ability to construct a visual reality is affected. The process for this is that an image is first projected onto the retina and the brain then interprets the meaning of the scene according to the visual input from the retina. Following this, the brain then utilises both emotional and rational intelligence to construct the reality of what is being seen.

For Damasio (1994, 1999) and Hoffman (1998), visual imagery is built on experiences, and is constructed and manipulated by the brain. As the foundation for thought, it is also considered to be the foundation for creativity. Finally, they conclude that visual experience is the "raw material" (Hoffman, 1998, p. 202) in all imagined constructions. Thus a designer can experience, see and read and thereby collect the necessary information to create a library of mental images. It is this ability that allows the designer to call upon mental images causing a dialogue to occur in the process of solving a design problem and creating of a novel solution to a design problem.

Piaget's (1971) research suggests that perception, mental imagery and language develop hierarchically, and form the basis of thought. According to his work, perception and mental imagery share a common basis in activity. He describes how mental imagery develops through action and can be developed through activities that involve imitation. Perceptual development is based on activity while the imitation of things, particularly through drawing, is linked to the development of mental imagery.

In his research on mental imagery, Piaget distinguished three forms of visual images, which he called static, kinetic and transformational images. Static images represent objects which do not rotate or move; kinetic images are those images which are based on the experience of an object's movement; and transformational images are those in which as an object is viewed it changes shape or form rather than position. This requires the transformation of a mental representation of the object, through a process involving mental manipulation of the visual image. Whilst Piaget's theories were based on extensive observation of children, he suggests that perception, mental imagery and language develop over the whole life span of an individual, as separate processes which are used in thought (Piaget, 1971).

The ability to work with static, kinetic and transformational images and to manipulate these mentally is fundamental to a designer; in particular, these different forms of mental imagery are central to the design process, and form the basis of the ways in which designers represent and communicate their ideas through sketching and drawing. Indeed there is growing research on the relationship between sketching and imaging. Sketching is one of the most common tangible activities in many fields of design and is a means by which the images in the mind's eye can be externalised by the designer.

The work of Goldschmidt (1983) was previously discussed in relation to the development of a model of designing. Wanting to give greater emphasis to design as a cognitive activity, Goldschmidt (2004) draws on the work of Fish and Scrivener (1990) to study the use of sketching in the design process, finding what she calls an *oscillation* between propositional thinking and descriptive thinking. One experiment was the observation of designers using two types of argumentation, based on these two types of thinking, during the process of design. The first is described as depictive and sketch based; and the second is non-figural. The use of these two types of argumentation oscillates over much iteration, particularly in skilled designers. Goldschmidt (2004) states that: "the order in which arguments switch modalities is not important. What is significant is the fact that the shift occurs both ways.... [It] helps translate the particulars of form into generic qualities, and generic rules into specific appearances" (Goldschmidt, 1991, pp. 138-139). In non-sketching protocols, Goldschmidt reports that "'seeing as' arguments are by far in the minority" (Goldschmidt, 2004, p. 140).

In their research, Fish and Scrivener (1990) describe the interplay between the two modes of mental representation – propositional (largely symbolic) and analogue ("quasipictorial, spatially depictive") – required for sketching (Fish & Scrivener, 1990, p. 121). As designers sketch, what they are seeing always undergoes a form of mental manipulation before it is represented as marks; in turn, the marks "generate mental images that may in turn influence the sketch" (Fish & Scrivener, 1990, p. 120). Although these authors state that in the activity of sketching, designers pull images directly from real life situations and objects for the most part, they also state that "there is objective evidence that spatially depictive images generated from memory have many of the properties needed to explain the ability of artists and designers to generate, manipulate, combine and inspect in imagination non-existent visual objects" (Fish & Scrivener, 1990, p. 122).

Although Laseau (1986) did not conduct formal studies in the area of sketching in architecture, he brought several decades of observational and professional experience to bear on his discussions of visual artefacts that represent the non-formal properties of a design space. His work makes it clear that visual artefacts are not produced solely for the sake of picturing an essentially visual product. To highlight these experiential insights, Suwa and Tversky (1996), in their study of sketching among architecture students and professionals, conclude that study participants were able to use their sketches to explore not only visual relationships among parts but also to explore functional relationships (such as lighting or circulation). They note that: "This analysis has revealed that sketches make apparent not only perceptual relations but also inherently non-visual functional relations to both advanced design students and practicing architects" (Suwa & Tversky, 1996, p. 192).

Drawing upon Goodman's (1976) taxonomy of symbol systems in which non-notational systems are defined as ambiguous with respect to what a given mark may stand for, Goel (1995) offers an explanation for the mechanism whereby sketching supports cognition during the design process. They indicate that sketches leave options open because the elements of the symbol system used in sketching are non-notational. As an example, a circle in a sketch may represent a sun, a wheel, a plate, or anything round. This ambiguity allows the designer to defer specific interpretation of the sketch (or diagram, as described by Laseau, 1986) and entertain or realise alternatives within the image that may not have been intended at the time of its creation. Supporting the neurophysiologic view of representations in the brain as quasi-pictorial and manipulable, they assert that "sketch indeterminacy may trigger innate recognition search mechanisms into generating a stream of imagery useful to invention" (Fish & Scrivener, 1990, pp. 121-123).

Examining the process of mental imaging in the context of computerised visualisation and industrial design, Athavankar (1996, 1997, 1999) raises some issues pertinent to design education. Whilst mental imaging differs between industrial design (typically small-scale objects of use) and architecture and interior design (larger-scale spaces which people typically inhabit), Athavankar does raise an important question:

> Design education traditionally has emphasized sketching and its use in creative explorations and is now supporting the use of computers as an alternative representation tool. The education has neglected the development of visualization and imaging abilities, not fully realizing their potentials as well as implications for creative pursuits. There are no conscious attempts to promote the development of imagery and abilities to control images voluntarily and thereby facilitate problem solving. Knowing full well that working on computer workstations makes greater demands on imaging abilities, how effective will future designers be, without the ability to develop virtual models in their mind to support their thinking? (Athavankar, 1997, p. 39)

Athavankar (1997) is one of few design theorists who specifically address the notion of imagining in designing as its own process, unrelated to sketching or decision making. Folkmann (2010) is one such theorist who specifically describes and analyses the process of imagination, seeking structural features of the imagination in the dynamic interaction between consciousness and the exterior material work, and how that relates to the design process.

Folkmann (2010) argues that imagining in the design process has received little focus in design theory, despite there being several studies of significance (Folkmann, 2010, p. 1). One such study is by Liddament (2000), who argues that not only is it impossible to know what is going on in the inner space of consciousness with regard to imaging and picturing in connection with design ('what goes on in the designer's head'), the very notion of a particular essence of creative imagination is problematic (Liddament 2000). In his seminal article, "The myths of imagery", Liddament (2000) criticises the postulation that in producing pictures, we render explicit something that already exists inside us as a sort of essence. He also criticises empirical approaches in cognitive science, in similar pursuit of a particular mental substance. For Liddement (2000), 'imaging' and 'imagery' is "not something intangible which takes place in a mysterious 'medium'" (for example, in the mind), but instead "imaging is a doing" that "alludes to the thinkable, and this means: to the do-able" (Liddement, 2000, p. 604).

Moore (2003), in a discussion on the role of visualisation in design, argues that it is important to be critical to the implicit 'metaphysical' notion that lurks in theories of creativity, especially critical of specific types of visual or sensory modes of thinking whose supposed roots are found in elements of consciousness that precede perception and language (Moore, 2003). She argues that there "is no need to look for anything hidden beyond or beneath what is already there in front of our eyes," (Moore, 2003, p. 12) and in an attempt to "demystify the art of design" she advocates a non-circumventable and opaque role of the visual in design.

Folkmann (2010) agrees, stating that "it is necessary to be critical of any metaphysical assumptions in the concepts we employ when speaking about design, imagination, and creativity. Still, it may prove productive for a discourse on creativity to address some features of the modus operandi of imagination in design" (Folkmann, 2010, p. 1). He perceives that imagination is a "structure that comes to itself in the dynamic interaction between inside and outside" (Folkmann, 2010, p. 2), in much the same way that cognition in design should not be regarded (only) as thinking but rather as an activity of inquiry and action that is flexible due to its specific function.

Within this dynamic interaction, designers construct meaning between mental settings and physical manifestations in design, which Folkmann (2010) proposes is "schematization", a concept that "captures the cognitive, imaginative framing of reality" (Folkmann, 2010, p. 2). Taking its roots from design discourse, the term schemata has been used to describe dominant ways of addressing problem solving in the "development of a growing pool of precedent" (Lawson 2004, p. 456).

This concept of 'schematization' also links with the idea that design is hemeneutical and the act of designing is one of 'positioning' (Snodgrass & Coyne, 2006).

Folkmann (2010) proposes that there are "three general meta-conceptual concepts or settings that are effective in the design process of turning inner imaginings into products" (Folkmann, 2010, p. 3). These include: first, 1) presupposed knowledge: known versus unknown; 2) Imaginative starting point: whole versus detail; and 3) degree of focus: focussing versus defocussing. Essentially, known versus unknown involves integrating layers of meaning that are unknown, emergent and becoming, and a mental setting "embraces the openness of the interface between known and unknown may make it possible to let the inner space of imaginings develop into something new in the design process" (Folkmann, 2010, p. 4). The whole versus detail is that, in design, the whole and detail are inextricably linked; "every detail is structurally and hermeneutically bound to a totality or a whole that can perhaps hardly be fully stated. Thus, details can only be developed and understood as fragments in the light of a totality that is perhaps only on the verge of becoming through fragmented details" (Folkmann, 2010, p. 5). In terms of focussing versus defocussing, considered in the context of imagination, Folkmann (2010) explains that:

> The relationship between problem statement and solution generation defines the path from inner imagining as an adaptation of the problem statement to outward manifestation in a design solution. Thus, the attention shifts from framing as a discursive activity of naming to focussing as in the process of schematizing a way of structuring the transfer of meaning between inner imaginings and outer physical manifestations" (Folkmann, 2010, p. 7).

A process that is similar to framing, Folkmann (2010) sees that the process of focussing lies in the interface between inner consciousness and outer world.

The notion of focussing and defocussing is also referred to by Kavakli and Gero (2001), who discuss 'defocussed attention' or 'remote association' as a method of "divergent thinking which refers to the general process of thinking of unusual associations", thus it is "important to deliberately defocus one's attention when attempting to discover creative solutions to a problem" (Kavakli & Gero, 2001, pp. 358-9). Both Kavakli and Gero (2001), and Folkmann (2010) agree that "Absolute focussing and defocussing cannot, however, be attained simultaneously. Instead, focussing and defocussing can be present in various degrees at the same time, or a design process may involve variations in focussing strategies" (Folkmann, 2010, p. 7).

Mental models, simulation and imagining

In design and architecture, mental simulation has been anecdotally described as an activity of utmost importance. An example of one famous anecdote is where Tafel (1979) describes how Frank Lloyd Wright in 1928 developed the concept for Fallingwater, an iconic architectural residential building in the United States. The building was commissioned by Edgar Kaufmann, who kept in contact via telephone with Wright to enquire how the plans were progressing. For almost one and a half years, Wright's response was simply that the plans were proceeding well, although no drawings were undertaken. One day, Kaufmann called and proclaimed that he was en route, two hours and twenty minutes away. It was only at that point that Wright began sketching plans for Fallingwater; first and second floor plans, with sections, elevations and details. All were drawn up almost true to final form, apparently developed fully in Wright's head prior to producing external representations. It was later that Wright described his design process in the following manner:

> Conceive the building in the imagination, not on paper but the mind, thoroughly – before touching paper. . . Let it live there – gradually taking more definite form before committing it to the drafting board. When the thing lives for you – start to plan it with tools. Not before. To draw during conception or sketch, as we say, experimenting with practical adjustments to scale is well enough if the conception is clear enough to be firmly held...But if the original concept is lost as the drawing proceeds, throw all away and begin afresh. (Wright, 1928)

This reflection suggests that design concept development is able to be carried out by using only mental simulation and imagery, and that the concept perhaps should not be committed to external representations (paper or prototype) prior to a process of consolidation. This aligns with the mental model book proposed by Gentner (2002), where a mental model is a representation of some domain or situation that supports understanding, reasoning and prediction. Supporting this book, Craik (1943) was the first to propose a theory of mental models as dynamic representations or simulations of the world; since then, however, the 'mental models' research paradigm has not evolved substantially (Forbus & Gentner, 1997; Markman & Gentner, 2001). While there has been a number of different theories, each relies on different assumptions of the nature of the cognitive system and focus their research on somewhat different aspects of mental models.

Theories on mental models illustrate that they represent different interpretations of the possible relationship among objects and their properties based upon given information, with one approach focusing on how people perform logical reasoning tasks with mental models (Johnson-Laird, 1983). The other approach is centred on causal mental models that are used in reasoning, which are based on long-term domain knowledge or theories (Gentner & Stevens, 1983). Several researchers have found that causal mental models rely on qualitative relationships, such as signs and ordinal relationships (Forbus, 1984; Kuipers, 1994). However, more important to this research is the fact that when forming mental models, people typically do not estimate things exactly as they would actually be, but rather a partial knowledge of the end outcome, as in the case of a designer who may have a partial knowledge of the space that they are to design.

An important feature of causal mental models is that they often allow mental simulation: being able to dynamically 'run' a simulation internally to observe functioning and outcome of a system or device. This ability to simulate something unknown means that an individual can predict outcomes even for situations where they have had no previous experience. The potential advantages of using a mental model in design includes the ability to reason about how physical spaces will allow for interaction under various circumstances with altered aspects without having to resort to physically constructing such a space. This ability is particularly constructive in other typically non- creative domains, such as science, as well as art and design, where uncertainty is an inescapable part of the problem space.

Essentially, one of the key components to mental simulation theories is the assumption that mental simulations help gain knowledge of the actual world through simulations of multiple possible (either past or future) alternatives (Forbus, 1984); knowledge is gained mainly through inferences (Gentner, 2002), leading to the generation of possible alternatives or predictions for future encounters with similar situations. Functionally, mental simulations may thus reduce information uncertainty.

Mental simulations are entirely subjective, operating with inexact or missing information, leading to partial, imprecise or approximate results (Forbus et al., 1997; Kuipers, 1994). Furthermore, although mental simulation can be said to be biased and prone to error in some respects, it is still a potent reasoning strategy allowing for the quick and cheap generation of approximate or imprecise knowledge in situations where the production of exact quantitative simulations is inappropriate, unavailable or impossible, such as in the design process.

Important aspects of mental simulations are the external factors that evoke or influence the use of mental simulations; mental simulations do not occur alone or in a vacuum. At times, the individual, in aiming to solve the problem, undertakes everything mentally, but eventually, mental simulations are done in the context of a physical object (such as models), sketches or drawings. Parts of the object may be mentally modified or transformed, and externalisation such as modelling or sketching, serves as both a help (providing a concrete base situation) as well as a hindrance (interference between the perceived world and the imagined world).

According to Norman (in Gentner & Stevens, 1983), mental models are representations of reality that people use to understand specific phenomena, describing them as such: “In interacting with the environment, with others, and with the artifacts of technology, people form internal, mental models of themselves and of the things with which they are interacting. These models provide predictive and explanatory power for understanding the interaction” (Norman, in Gentner & Stevens, 1983, p. 55).

Mental models are consistent with theories that link internal representations with thinking processes. Johnson-Laird (1983) proposes mental models as the basic structure of cognition: “It is now plausible to suppose that mental models play a central and unifying role in representing objects, states of affairs, sequences of events, the way the world is, and the social and psychological actions of daily life" (Johnson-Laird, 1983, p. 397).

Adding to this, Holland, Holyoak, Nisbett & Thagard (1986) suggest that mental models are the basis for all reasoning processes: "Models are best understood as assemblages of synchronic and diachronic rules organized into default hierarchies and clustered into categories. The rules comprising the model act in accord with the principle of limited parallelism, both competing and supporting one another" (Holland, et al., 1986, p. 343).

Expressed another way, mental models are psychological representations of real, hypothetical, or imaginary situations. The mind constructs 'small-scale models' of reality that it uses to anticipate events, to reason, and to underlie explanation (Craik, 1943). These models have a structure that corresponds to the structure of what they represent. They are accordingly akin to architects' models of buildings, to molecular biologists' models of complex molecules, and to physicists' diagrams of particle interactions. Since Craik's (1943) original theory, researchers in cognitive science have argued that the mind constructs mental models as a result of perception, imagination and knowledge, and the understanding of dialogue.

Lawson (1980) argues that reasoning and imagining are most important to designers as a critical part of the design process. Whilst reasoning and imagining differ in that reasoning "is considered purposive and is directed toward a particular conclusion" and imagining is where the individual draws from their own experience "…combining material in a relatively unstructured and perhaps aimless way", both are considered to be a part of the creative process of designing. He explains that even in the most structured and disciplined fields such as engineering, many design problems are solved using the combination of imaging, imagining and reasoning in this creative and imaginative process (Lawson, 1980, pp. 137-138).

While there is a reasonable body of literature focussing on imaging and mental models, very little explicitly focuses on imagining, with the exception of Folkmann (2010), Moore (2003) and Liddement (2000). Other than these authors, where there are references to imagining as in the previous examples, they are broad and vague. Having said this, where imagining is used (and used as distinct from imaging) there seems to be an implicit understanding of it as a 'meta' activity embracing and being supported by other activities such as imaging and mental simulation. In this regard, it invites closer exploration of the relationship between imagining and the process of creative mental synthesis emerging in more recent research.

Creative mental synthesis

According to Dorst (2011), "Creative design seems more to be a matter of developing and refining together both the formulation of a problem and ideas for a solution…" (Dorst, 2011, p. 434) with constant iteration between, or as described by Maher, Poon and Boulanger (1996) 'co-evolution' of, the "problem space" and the "solution space" (both terms coined by Newell and Simon, 1972). The process of iteration is in effect bridge building facilitated by the identification of a key concept or 'frame'. "The ability to frame a problematic situation in new and interesting ways is widely seen as one of the key characteristics of design thinking" (Paton & Dorst, 2011, p. 1). Where possible, it is also seen as increasingly important for this process to be undertaken collaboratively with the client and, when necessary, to enable negotiation and reframing. In research by Paton and Dorst (2011) this involved the use of metaphor and analogy, contextual engagement, and conjecture (Paton & Dorst, 2011, p. 8). Contextual engagement or immersion is regarded here as "a means not only to learn about the design situation, but also to strategically understand how to create meaningful nodes of interaction and helpful activities to facilitate reframing with a particular client" (Paton & Dorst, 2011, p. 9).

According to Kolko (2010), "Synthesis is an abductive sensemaking process. Through efforts of data manipulation, organization, pruning, and filtering, designers produce information and knowledge" (Kolko, 2010, p. 17), and abduction allows for the creation of new knowledge and insight in the complexities that the designer may have to deal with during the design process. For Kolko (2010), abduction is "the hypothesis that makes the most sense given observed phenomenon or data and based on prior experience" (Kolko, 2010, p. 3). Within the industrial design context, Kolko (2010) describes sensemaking as an attempt to understand connections and their implications for future action. Invariably because of the huge amount of data, this involves as part of the synthesis process a 'getting it out' or externalisation of the collected information. Incorporated with this is a process of spatialisation, and the development of a mental model of the design space to be explored and refined. In this respect, Kolko (2010) advocates externalising the entire meaning-creation process; that is, removing data from the cognitive realm (the brain) as well as from the digital realm (the computer), to the physical realm, such as a wall where the designer (and the client) can see the whole problem space and is able to prioritise, judge and forge connections between data.

The concept of 'framing' provides further clarification and as previously described is understood to be central to the designer prioritising, judging and forging connections. For Schön (1984), a design hypothesis "depends on a normative framing of the situation, a setting of some problems to be solved" (Schön, 1984, pp. 134-136). This usually involves a person, a setting, or an action-based goal where the designer will 'position' self within the space or place (mentally) and 'enact' through imagining a similar scenario, and 'experience' certain aspects of the space or place to be designed.

During the process of design, it is synthesis where cohesion is revealed and clarity ensues, yet it is also a process that appears almost 'magical' to the outsider, as synthesis is frequently performed privately (in the head, or mind's eye), contrary to other aspects of the design process (such as sketching and drawing) which are visible to non-designers. This stage of designing is critical and is a considered to be an insular process, one where it is practised as a private exercise and "there is no visible connection between the input and the output; often, even the designers themselves are unable to articulate exactly why their design insights are valuable" (Kolko, 2010, p. 15). As an observer, this is considered to be a 'magical process'; one where there is no apparent link between the beginning and end process of designing. From the designer's perspective it is a deep and reflective process, and is difficult to articulate. This gap in understanding leads to an ongoing problem in design where design research and design synthesis are 'de-valued' as both are considered to be an 'informal' step in the design process. Kolko (2010) articulates it where "The design output and solutions can be unique, novel, and even exciting, but because there is no artifact-based procedural trail, the client isn't aware of the various internal deliberations that have occurred" (Kolko, 2010, p. 17).

As current research reveals, non-design areas are showing interest in the process of creative mental synthesis. At the 8th Design Thinking Research Symposium in Sydney, Australia in 2011, there was growing interest in design thinking outside the design areas, attributed in part to a changing world and associated wicked problems. In addition to its designerly strategies for dealing with wicked problems, design is also recognised for its ability to affect change through the artefacts which it produces (Stewart, 2011, p. 2). In reviewing the papers presented at the Symposium, Stewart (2011) observes that the common thread "is an interest in design as an interpretive practice within which particular kinds of sense-making are operative" (Stewart, 2011, pp. 4-5).

Originally used as a concept to describe designers' cognitive strategies of problem solving in the design disciplines, such as in research by Cross (1982), Lawson (1980), Nagai and Noguchi (2003) and Papantonopoulos (2004), design thinking is emerging as a concept marketed to advance innovation in non-design areas (Brown, T., 2008; Grots & Pratschke, 2009). The concept of design thinking has also found its way into academic curricula well beyond traditional design programs; for instance, the Rotman School of Management (Toronto) in the context of the Masters of Business Administration (MBA) education, or at Stanford and Potsdam which offer design thinking education intended specifically for non-designers (Dunn & Martin, 2006; Plattner et al., 2009).

This area of design thinking is premised upon the notion that those leading a group of people should consider 'thinking like designers' (Dunne & Martin, 2006) or adopt a 'design attitude' (Boland & Collopy, 2004); thus, organisations should organise themselves like 'design teams' (Dunne & Martin, 2006). The premise of this is that by learning to 'think like a designer' one can transform the way that products are developed (this should not be confused with industrial design) the way that services are provided and how processes and strategies are carried out. Furthermore, design thinking is seen to play a key role in innovation (UK Design Council, 2009).

Summary and discussion

This chapter presented an overview of design methodology research focussing on design process, design thinking and cognition, and imagining. In terms of design process, the review pointed to early research concerned with developing prescriptive models of design in order to improve process efficiency and product performance. Underlying this was an understanding of design as a mechanistic, sequential activity. When these new models failed to achieve the desired outcomes, especially in the spatial design areas such as architecture, attention turned to better understanding the nature of the design task, and with this a new conception of designing as a heuristic, satisficing activity emerged. As the review shows, this was facilitated through research concerned with the nature of design problems and design as a (creative) cognitive activity. Design problems were identified as being ill-defined and 'wicked' due to their complexity and future-oriented nature demanding a generative way of reasoning involving abductive, as well as inductive and deductive, thinking. In association, increasing attention was given to imaging, visualisation and mental imaging, and more specifically, the role of mental models and simulation. While the review revealed few specific references to imagining, recent research on creative mental synthesis appears to provide a fertile basis for ongoing research of relevance to design including the spatial design areas.

As Kokotovich (2000) identified, although much research has been undertaken in the area, there is still a need to use a more systematic methodology in order to extend the views expressed in the design literature relating to creative mental synthesis. He argues that whilst there has been extensive research over the past thirty years in perceptual psychology, the research has not specifically addressed issues in design thinking. As such, he proposes that "[D]eveloping a detailed understanding of creative mental synthesis will serve to support design education, and therefore the improvement of design practice. Cognitive processes are central to the process and practice of design. Consequently, it is important that some of these cognitive processes be identified and understood" (Kokotovich, 2000, p. 2).

Such a recommendation is relevant today with Paton and Dorst (2011) proposing that design education "would benefit from making reframing and briefing strategies salient for students: as a process at the project level; and, to foster expertise in creating new frames at the professional level" (Paton & Dorst, 2011, p. 14). One of their recommendations for future research is "investigating the professional meta-activities designers engage in to support fostering new frames; and understanding the different ways a frame can change" (Paton & Dorst, 2011, p. 14).

However, even with significant development over the past five to six decades, design research remains in a state of confusion (Visser, 2006a, 2006b, 2006c, 2009, 2010b). Over the past few decades, a number of movements in design methodology and research have been introduced, as discussed in depth above. These movements in design thinking and theorising have transformed over the decades, developing into various discourses and even splitting into several different directions which have very little to do with each other. Visser (2006a) argues that, to date, no significant leading theory and methodology has emerged, and this feeling is echoed by industrial design theorist Krippendorff (2006), who believes that old theories and methodologies have faded, and what remains in design discourse is an odd mixture of various older style approaches. Even more disappointing is the dearth of research in the spatial design areas (such as architecture and interior design) reinforcing the overgeneralised view that design research undertaken in industrial design is also relevant for the spatial design disciplines. The research described in this book is an attempt to respond to these issues.

Chapter 3: The Presence Research Framework

Introduction

As conveyed previously, the research described in this book commenced with an exclusive focus on imagining in design. After initial exploration of students' experience of designing and further reading in the area of design methodology, the lack of research concerning the cognitive process of imagining became evident. This prompted additional reading beyond the design disciplines and a subsequent connection to presence research and disciplines that have been significant in shaping it such as cognitive science, engineering and computer mediated communication (CMC).

Specifically, the content forming this chapter was structured through five main sections informed largely by Lombard and M.T. Jones (2007) (Figure 3.1).

The first section introduces the presence research framework and presents the historical foundations of presence and its relationship with cognition. In the second section, the various conceptualisations of technology mediated presence are discussed. The seminal texts regarding both non-mediated and non-technology mediated presence are examined in detail in section three, as are the notions of co-presence and spatial presence. In addressing the presence process, the 'book problem' is defined as a debate that centres upon the question of whether less immersive media are capable of providing a presence experience. The fourth section examines current research on the debate surrounding presence as external perception or internal conceptual processing, underlining the potential gaps in both aspects. The fifth section focusses on mental models, the concept of the suspension of disbelief and facilitating mental simulation. The sixth and final section highlights seminal issues and constructs of potential interest to imagining in spatial design.

This chapter, then, provides an overview of presence research including its history and areas of investigation that are of particular relevance to imagining, such as technology mediated presence, presence mediated by a non-technological external or internal element (such as a book or mental model), and non-mediated presence (Figure 3.1). As indicated in Figure 3.2, the review gives special attention to the role of cognition in presence research aligning with the explicit interest given in the book to the cognitive process of imagining In addition, it highlights in a concluding section the perceptual/conceptual debate that currently divides the presence research community.

Presence Research Framework

Traditionally, the term 'presence' has been defined as a state of being present in a place or something felt or believed to be present; it indicates either a tangible condition when something or someone is actually present in the physical world, or may also connote a personal perception of the world (physical or virtual), embodied in a feeling or belief.

It is defined as a subjective experience of 'being' and 'acting' in a virtual environment (Slater, Usoh & Steed, 1994), usually in the sense of being in a computer-generated or computer-mediated environment. Research suggests that the presence experience depends upon certain features of the virtual environment, for example, techniques and types of interaction (Regenbrecht & Schubert, 2002). It should be noted that in early references to presence, the term telepresence is used as originally coined by Minsky (1980) to reflect its association with technology. However, the more general term 'presence' is generally used in the presence research community.

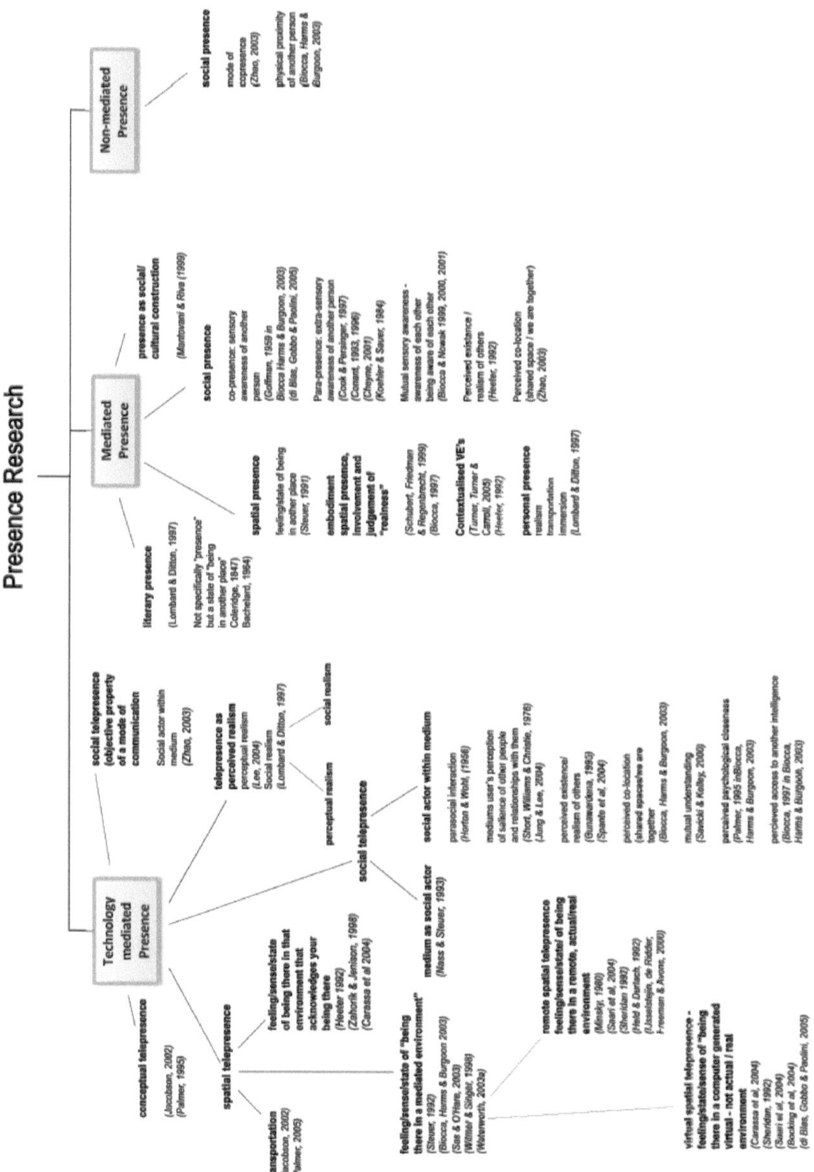

Figure 3.1: Literature review framework for presence research informed by Lombard & M.T. Jones, (2007).

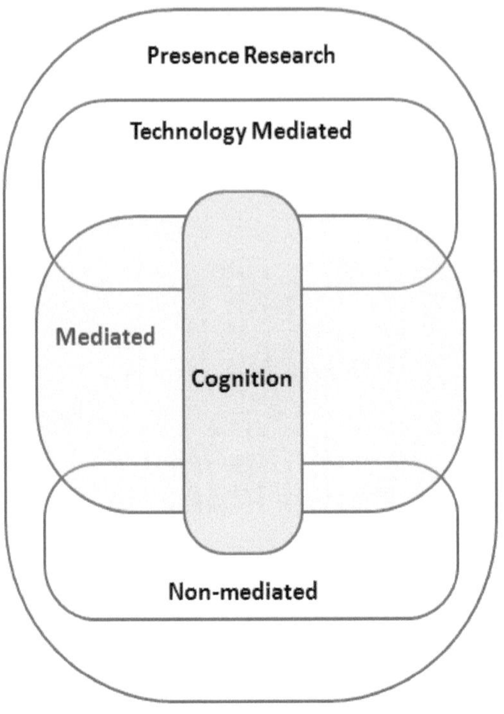

Figure 3.2 Literature review context for presence research

An overview of the historical aspects and highlights of presence research reveals that the first decade of research commencing in the mid to late 1990s was dominated by a comprehensive debate on the definition of the phenomenon of presence and an understanding of it as a psychological phenomenon that is experienced or felt (Lee, 2004; Lombard & Ditton, 1997; Lombard & Jones, 2007, 2009). Characterising and raising an awareness of presence meant that the experience is distinguished significantly from automatic and uncontrollable responses to virtual environments such as startle reflex, body sway, physiological reactions, or social responses, which have sometimes been discussed as alternative, objective measures of presence (Bailenson, Blascovich, Beall & Loomis, 2003; Freeman, Avons, Meddis, Pearson & IJsselsteijn, 2000; IJsselsteijn, de Ridder, Freeman & Avons, 2000; Slater et al., 2006). It is important to note that this was also a period in presence research where quantitative methods were the primary methods used in order to understand the phenomenon.

From approximately 2005 to the present, research on presence has gained new momentum and a new direction. In this more recent phase, researchers are recognising that cognitive theories of spatial presence are needed. Rather than focussing on *what is presence*, it is now the *how does presence occur* that is important with qualitative methods used in conjunction with quantitative methods (Garau, 2003; Slater et al., 2008; Thomsen, 2004; Turner & Turner, 2006).

Lee (2004) stresses that one of the most important ongoing concerns in presence research is to explain the "mental mechanism that enables humans to feel presence when they use media or simulated technologies" (Lee, 2004, p. 47). Theories are currently being developed that attempt to explain which cognitive processes are involved in the perception and interaction with virtual environments, and how these processes lead to presence (Lombard, 2011; Riva, 2011; Turner & Turner, 2011). In doing so, these theories are largely based on the proposition that unconscious spatial cognitive processes underlie the construction of a mental model of the virtual environment, which, in turn, evokes presence (Biocca, 1997, 2002, 2003; IJsselsteijn, 2002; Nunez, 2007; Regenbrecht & Schubert, 2002; Wirth, Hartmann, Bocking, Vorderer, Klimmt & Schramm, 2007).

Therefore, from the 'definition of presence' debate arising earlier in the field, combined with subsequent research on determinants and measurements of presence (Slater, Linakis, Usoh & Kooper, 1996; Slater & Usoh, 1993; Slater & Wilbur, 1997) and attempts to develop a deeper understanding about the *process* of the presence experience, a significant body of knowledge is developing. In spite of this, there appears to have been no concerted attempt to link the cognitive aspects of the design process (in particular imagining) with presence research.

Informed by Lombard and Jones (2007), this book proposes three main categories of presence: presence which is brought about only through technology (technology mediated presence), presence which occurs via a non technological external element (such as a book or an internally generated mental model), and finally, presence experienced when in actual physical proximity to another person or object (non-mediated physical presence) (Figure 3.3).

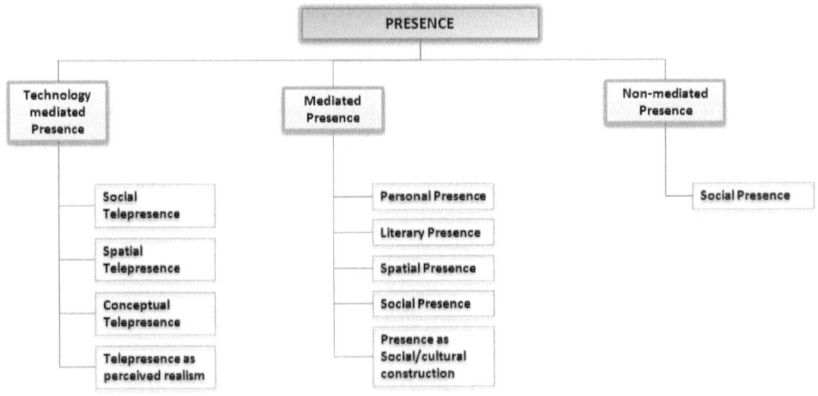

Figure 3.3: Presence categorisation adapted from Lombard and Jones, M. T., (2007)

The three categories conveyed in Figure 3.3 can be expanded further using Lombard and Ditton's (1997) six conceptions of presence: social richness (the 'warmth' or 'intimacy' possible via a medium), realism (perceptual and/or social), transportation (the sensations of 'you are there', 'it is here' and/or 'we are together'), immersion (in a mediated environment), as a social actor within medium (e.g., parasocial interaction), and the medium as social actor (e.g., treating computers as social entities).

Presence and cognition

Before exploring the three main categories of presence in detail and the role played by cognition, a brief discussion on the general relationship between presence and cognition is provided. Although many early studies acknowledged that presence is related in some way to cognition, a significant step in the correlation between presence and consciousness was proposed by Loomis (1992), creating a new disciplinary relationship between cognitive science and presence. However, current research within presence identifies that since Loomis' early work, the contribution that cognitive science provides to presence research is an ongoing issue that few agree upon. To date, Riva remains the most prolific author on the relationship between cognitive science and presence. Riva's (2009) studies within the area propose a new model in presence research where the environment (real or virtual) "...offers different opportunities and produces presence according to its ability in supporting the users and intentions" (Riva, 2009, p. 167).

Considering both mediated as well as non-mediated presence, Riva, Baccetta, Cesa, Conti & Molinari (2003) argue that the essence of presence research "is the comparison of human perceptions and responses in the context of technology with human perceptions and responses in contexts that do not involve technology" (Riva et al., 2003, p. 2). This is based upon the definition given by the International Society of Presence Research (ISPR), where 'presence' is a "psychological state in which even though part or all of an individual's current experience is generated by and/or filtered through human-made technology, part or all of the individual's perception fails to accurately acknowledge the role of the technology in the experience" (ISPR 2000).

Riva (2009) maintains that the recent outcomes of cognitive science offer a broader definition of presence that is not limited to technology where "presence is described here as a core neuropsychological phenomenon whose goal is to produce a sense of agency and control: subjects are "present" if they are able to enact in an external world their intentions" (Riva, 2009, p. 1; Riva et. al., 2006). Furthermore, the relationship between presence and action (agency) is linked through the subconscious separation of both 'internal' and 'external', and 'self' and 'other' (Riva, 2007). Presence is defined as the non-mediated perception of successfully transforming intentions in action within an external world, and social presence is defined as the non-mediated perception of an enacting other (recognising others' intentions) within an external world (Riva 2006, 2008a, 2008b). Put simply, the presence experience is based upon the individual experience. Additionally, the more complex the task, the more likely the sense of presence experienced. Finally, maximum presence is experienced when the environment is able to support the full intent of the user. That is, the more that an individual feels that he/she can 'act' (sense of agency) within an environment, the greater the sense of presence experienced.

To summarise, relating presence and agency offers a conceptual framework for gaining an understanding of the relationship involving action and recognition of intentions. Presence is seen as a basic human experience encompassing everything from subconscious physiological responses through to higher-level cognitive, emotional and bodily awareness, through physical and social interaction. As such, presence research, and its link with agency provides an enormous possibility for the exploration of environments allowing different opportunities for supporting users and their intentions. However, whilst the focus is on presence and agency, there are still opportunities to scrutinise different forms of presence encompassing perception, neurological processing, cognition, interaction and emotions, affecting people and culture, especially in terms of designing. Examining the different conceptualisations of presence may facilitate development of elements which capture previously undiscovered aspects of human presence and interaction.

Technology Mediated Presence

As outlined earlier in Figure 3.3 and highlighted here in Figure 3.4, there are several forms of technology mediated presence (telepresence): social telepresence, spatial telepresence; conceptual telepresence and telepresence as perceived realism.

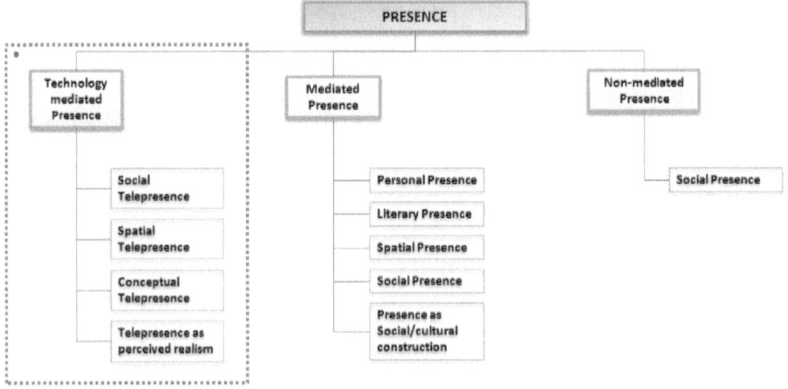

Figure 3.4: Technology mediated conceptualisations of presence

Technology mediated presence occurs where the mediating technology 'invisibly' enables a person to experience the perception of being somewhere other than where they are in reality. Minsky (1980) refers to this as telepresence. Lombard and Ditton (1997), who are cited widely in presence research literature, describe it as having the sense of 'being somewhere' usually in the sense of being in a computer-generated or computer-mediated environment. Further, they describe presence in terms of invisibility or transparency, as a large open window, with the medium user and medium content (objects and entities) sharing the same physical environment. "An illusion of non-mediation occurs when a person fails to perceive or acknowledge the existence of a medium in his/her communication environment and responds as he/she would if the medium were not there" (Lombard & Ditton, 1997, p. 5).

This type of presence is also referred to as 'virtual presence' which is defined as the creation of an illusion of presence created by artificial immersive devices; commonly affiliated with 'virtual reality' (VR) or 'virtual environment' (VE) systems, where computer programs are used to generate virtual objects and environments presented to the individual through a variety of technologies; artificial immersive input devices are used to stimulate the senses and to create an illusion of 'being there' within a remote virtual environment, thus resulting in the presence experience.

During the past three decades, there has been a significant focus on research and development of media that enhance the sense of presence as it provides more accurate reproductions and simulations of reality than were previously thought possible in immersive displays, computing and network technologies and interactive computer graphics. IJsselsteijn and Riva (2003) propose that "…research interest in presence has mainly been motivated by work in three related domains: teleoperation, simulation and telecommunication" and that "…presence research offers the possibility to engineer a better user experience, to optimize the effectivity [sic], efficiency and pleasurability of the different applications. From an application viewpoint, presence research will spur the development of numerous tele-applications in home and professional environments" (IJsselsteijn & Riva, 2003, p. 7).

Presence research not only benefits the entertainment industry, but has also advanced to become common 'currency' in areas such as virtual environments, advanced broadcast and cinematic displays, teleoperation systems and advanced telecommunication applications.

In terms of the subcategories for technology mediated presence, Lombard and Jones (2007) depict social telepresence as a social actor within a medium. Zhao (2003) describes this as a form of human co-location in which both individuals are present in person at their local sites but are located in each other's electronic proximity rather than physical proximity. Spatial telepresence is a feeling, a sense or a state of 'being there in a mediated environment'. Carassa, Morganti and

Tirassa (2004) propose that it is the subjective feeling or mental state in which an individual, through visual, auditory, or force displays generated by a computer, believes that they are 'physically present' within a virtual environment.

Conceptual telepresence differs from spatial telepresence in that it is the degree to which the individual feels involved with or absorbed in and engrossed by stimuli from the virtual environment (Palmer, 1995). Related to this is the concept of telepresence as perceived realism, which Lombard and Ditton (1997) describe as the extent to which a media portrayal is perceived as being plausible or 'true to life' in that it reflects events that do or could occur in the non-mediated (real) world. Lee (2004) describes this as the "psychological state in which virtual (para-authentic or artificial) objects are experienced as actual objects in either sensory or nonsensory ways" (Lee, 2004, p. 37).

Non-mediated (physical) presence and (non-technology) mediated presence

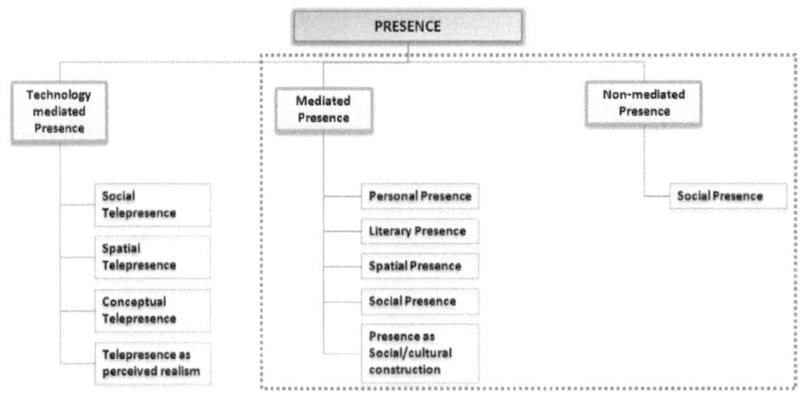

Figure 3.5: Mediated and non-mediated conceptualisations of presence.

Non-mediated physical presence is a sensation of existence with and/or in an objective actual something; or 'corporeal togetherness' as defined by Zhao (2003). According to Zhao (2003), corporeal togetherness explicitly excludes technology; it is "the most primitive mode of human togetherness" (Zhao, 2003, p. 447). To interact with someone in corporeal co-presence is to interact with that person face to face. It is introduced here with (non-technology) mediated presence (Figure 3.5) because social presence can be an aspect of mediated presence and also because people who experience mediated presence are co-present corporeally with a physical reality. This sense of social presence or corporeal co-presence also occurs in design, where people engage iteratively with internal and external worlds, experiencing and intentionally orchestrating the relationship between non-mediated (physical) presence and (non-technology) mediated presence.

Non-technology mediated presence is used to describe human experience in traditionally non-immersive and non-interactive environments. It is discussed in this section in order to highlight the various aspects of presence elicited by media that is non-technological; that is, any activity that elicits presence through non-digital means, such as reading books, dreaming or daydreaming. This form of presence is important to discuss given its potential connection to imagining in the design process.

As noted in the previous section, historically, the research on presence has been on powerful new media technologies which are able to create an illusion of 'being there' (for example, head mounted displays, CAVESs and VEs). In contrast, this section deals with presence that is considered to be experience based on external (or perceived as external) and/or internally generated stimuli. Where technology mediated presence describes the sense of presence in a virtual environment for a period of time, or an individual's interaction with virtual environment's entities, mediated/non-mediated presence could be best described as the situation where the individual fails to recognise the existence of a medium (telephone, book, or the mental imagery space). The stimuli which allow presence to occur can be either perceptual or conceptual, or even both.

As shown in Figure 3.5, the five sub-categories of non-technology mediated presence are personal presence, literary presence, spatial presence, social presence, and presence as social or cultural construction. Personal presence is the extent to which the individual feels a part of a virtual world, such as seeing the partial or whole representation of self. Literary presence can be defined as an illusion of presence created by storytelling, and this illusion is a common artistic goal for such traditional communication media as books, theatre, television, and film (Lombard & Ditton, 1997). Through text or narrative, the spoken voice, and images on film, an individual can be led to believe that they are somewhere they are not, or in the presence of people and objects that do not actually exist. Schubert (2009) defines spatial presence as "the *sense of being there* that applies to the experiences of spatial presence in virtual and real environments as well as to the experiences emerging from reading and remembering...irrespective of whether technology mediated the experience, or which technology was involved" (Schubert, 2009, p. 162). The terms 'physical presence', 'a sense of physical space', 'perceptual immersion', 'transportation', and the 'sense of being there' are all terms used interchangeably with spatial presence. Social presence, according to Heeter (1992), is the perceived existence of others: "the extent to which other beings (living or synthetic) also exist in the world and appear to react to you" (Heeter, 2009, p. 262).

The sub-category of presence as social or cultural construction stems from a 'cultural perspective', where Mantovani and Riva (1999) argue that 'reality' is not 'out there' in the world, somewhere 'outside' people's minds, escaping social negotiation and cultural mediation. Rather, reality is co-constructed in the relationship between people and their environments through the mediation of the artefacts. Further, they argue that this concept of presence expands the ecological approach in presence research that neglects the social and cultural dimension of experience, and that it recognises that experience is culturally mediated and immersed in a social context. The concept of presence as social or cultural construction is broken down into the following formulas:

- Presence is always mediated by both physical and conceptual tools that belong to a given culture
- The criterion for presence does not consist of simply reproducing the conditions of physical presence but in constructing environments in which actors may function in an ecologically valid way
- Action is essentially social (as knowledge in everyday situations is often distributed among various actors and various artefacts) (Mantovani & Riva, 1999)

As discussed above, Lombard and Ditton (1997) define presence as the psychological sense of 'being there'. Related to this is co-presence, which is the psychological sense of 'being together' in such an environment and can be defined as a form of human co-location where the participants can see each other. Associated with this can be what Lombard and Ditton (1997), in their seminal text on presence, describe as 'the perceptual illusion of non-mediation', produced by means of the disappearance of the medium from the conscious attention of an individual. They explain this as "when a person fails to perceive or acknowledge the existence of a medium in his/her communication environment and responds as he/she would if the medium were not there" (Lombard & Ditton, 1997, p. 9).

The concept of non-mediated presence as the experience of 'being there' in 'another world' which is not necessarily mediated by technology can be associated with the 'willing suspension of disbelief' identified by Coleridge (1847); 'reverie' identified by Bachelard (1971); and 'flow' identified by Csikszentmihalyi (1990).

In terms of literary presence, the experience of reading a highly engaging novel has the ability to mentally transport an individual from the physical environment in which they are located to the environment described in the text; an individual can be totally absorbed in this experience. Building on this spatial metaphor, Green and Brock (2000) describe this phenomenon as 'transportation'. Although Gerrig (1994) originally coined the term, Green and Brock's (2000) conceptual definition refers to a 'transporting effect' or immersion into a narrative.

In their description of transportation in fictional texts, they emphasise the role of attention: "we conceived of transportation as a convergent process, where all of the person's mental systems and capacities become focused on the events occurring in the narrative" (Green & Brock, 2000, pp. 701-702). They assume that mental imagery evoked by a story has an impact on the attitudes of the reader when it is activated in the state of high transportation, because transportation inhibits a critical scrutinising of the content and the 'message' of the imagery. Transportation is measured in various studies with self-report items focusing on the construction of vivid imagery, awareness of real environment, and affective involvement in the story.

Schubert and Crusius (2002) take a somewhat different stance, proposing that the sense of presence in a non-technology mediated environment is neither more nor less, but rather, is equal to presence experienced from means of technology. They suggest there is a need to acknowledge cognition as a mediator between immersion and presence, asserting that the fact that presence can emerge from both the perception of stimuli and the understanding of symbols indicates that the perceiving mental representations play a role in constructing the world in which we feel present. Schubert and Crusius (2002) assert that presence is not a direct result of immersion, but is "mediated by cognitive representations that are constructed on the basis of immersive stimuli" and that it is the "structure of this mental model [that] determines whether the user experiences a sense of presence or not" (Schubert & Crusius, 2002, p. 1).

In comparing the definitions and descriptions of presence, it can be seen that there are both similarities, as well as differences, between the various forms of presence in different media. Essentially, the core aspect of the presence experience is the metaphor of *travelling to another place*. However, where the traditional engineering paradigm of presence in virtual reality environments appears to be related to a sense of *actually being there*, the sense of transportation in literature appears most strongly related to *being absorbed* by the text, in various depths. Ryan (2001) makes explicit reference to presence as an emergent property of reading, arguing that four distinct degrees of absorption are distinguished. These are: concentration, which is when the information is complex and the reader concentrates too much to be immersed (usually in the case of complex storylines); imaginative involvement which is "The split subject attitude of the reader who transports herself into the textual world but remains able to contemplate it with aesthetic or epistemological detachment" (Ryan, 2001, p. 10); entrancement, described as total immersion where the reader is caught up in the textual world so that the surrounding world fades away and the reader feels as though they are taken to the 'world in the story'; and finally, addiction, which is the attitude and the willingness of the reader to escape from reality and being able to find a home in the textual world. It is also the condition in which a reader is so much in the grip of a textual or narrative world that they may feel compelled to interfere with certain aspects of it and experiences it not as if it were life (as a whole) but rather, as a 'slice' of life.

Put simply, alongside these levels of absorption, Ryan identifies that when in the grip of text or a narrative, the individual will experience spatial, temporal and emotion immersion; transportation to the geography of the narrated world; engagement with the unfolding events of the story; and identification with the characters portrayed.

Based upon the notion of transportation, Schubert and Crusius (2002) propose several theses, arguing that the psychological phenomenon of presence is the same in relation to three different 'tools' used to elicit presence: a virtual reality laboratory, a cinema, and a reading chair, and in all three media, the actual physical environment (VR laboratory, cinema, reading chair) is suppressed in favour of an alternative, medially presented and cognitively construed environment. The authors also assert that in order to understand the book problem (discussed later in this section), there remains the necessity to acknowledge the role of cognition as a mediator between immersion and presence: "The fact that presence can emerge both from the perception of visual stimuli and the understanding of symbols shows the necessity of another layer in a theoretical model of presence, namely that of mental representations" (Schubert & Crusius, 2002, p. 4).

This notion is also supported by Schubert, Friedmann and Regenbrecht (2001), who suggest that although users report presence in a virtual environment or presence in a fictional narrative, this experience is mediated by mentally constructing an environment surrounding the body. Furthermore, Schubert and Crusius (2002) state that presence in virtual reality, film and text all diverge with regard to the amount of spatial presence and involvement, and books can produce presence because they use the power of narration. Literature requires narration in order to produce a sense of transportation, and film strongly profits from it, although film can do without a narrative (Burch, 1990). The sense of 'being there' can also occur in readers of texts, because readers can imagine very vividly being present in the world that the narrative of the book creates. More precisely, readers may construct a mental model of their bodies and of possible actions in the environment described by the book (Biocca, 2003).

Glenberg (1997) integrates mental models and possible actions within the 'embodied cognition' framework. Mental representations of mediated spaces, which can include assumptions of what kinds of actions are possible in a given space are required for spatial presence, and in most cases, spatial presence is an experience which can be enriched by (but does not completely depend on) external (media-based) information such as visual, auditory, haptic, or proprioceptive impulses and feedbacks (Gibson, 1973; Kebeck, 1997). Proprioception is the sense of the orientation of one's limbs in space. This is distinct from the sense of balance, which derives from the fluids in the inner ear, and is called *equilibrioception*. The more senses a media environment activates in its users the more likely it is that the receivers will feel like they 'are' in the environment.

One argument for the capacity of books to elicit a sense of presence is the superb aesthetic experience of literary texts, which can include incredibly detailed and vivid portrayals of spatial configurations. Several examples of this include the stories of Marco Polo, the works of J. R. R. Tolkien, J. K. Rowling and C. S. Lewis.

Gysbers, Klimmt, Hartmann, Nosber and Vorderer (2004) argue that if spatial presence is known to occur in users of non-immersive media of books, direct sensory experience cannot be the only medium of presence and "higher-order mental activities such as cognitive involvement and imagination" are important for the facilitation of spatial presence (Gysbers et al., 2004, p. 13).

Advancing the notion of spatial presence is the conceptual model proposed by Vorderer, Wirth, Saari, Gouveia, Biocca, Jäncke, Böcking, Hartmann, Klimmt, Schramm, Laarni, Ravaja, Gouveia, Rebeiro, Sacau, Baumgartner, and Jäncke (2003) who propose that the presence experience occurs as a two-step process and can be experienced across various media but more importantly, including during reading.

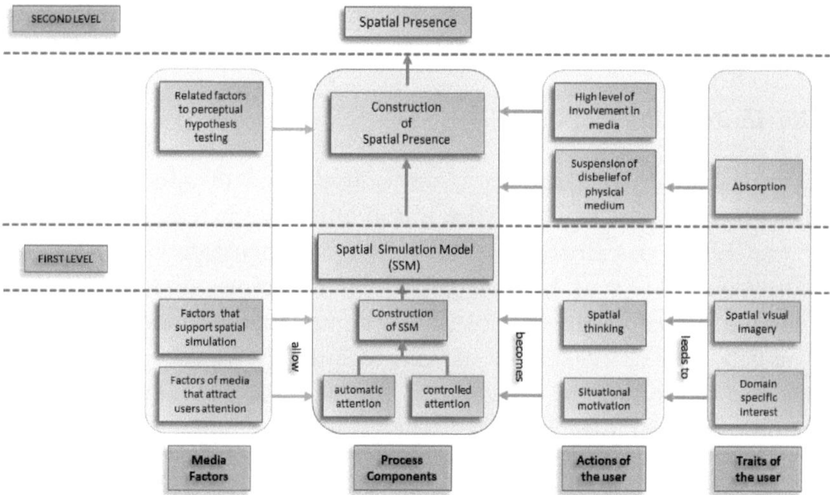

Figure 3.6: The MEC Model of Spatial Presence adapted from Vorderer et al. (2004)

The authors propose that "readers generate a mental representation of the spatial environment portrayed in the text" (Vorderer et al., 2004, p. 13) and that they do so by combining text-based knowledge and knowledge-based information in order to create a spatial situation model (SSM) of the environment described in the text. This process is explained as a method where an individual processes space-related information included within the text (which is the bottom-up component) and by adding spatial images with space-related knowledge already available in the mind before exposure to the text (the top-down component). Put simply, readers combine text-based and knowledge-based information to create a SSM. This is illustrated in Figure 3.6 as part of the MEC model of spatial presence.

However, spatial presence only occurs if the individual considers self to be actually located within the SSM and no longer believes self to be part of the real environment. Therefore, it is not enough to simply imagine how lovely an environment would look like, but rather, the person must actually regard themselves as physically there within that environment.

Based upon this model, experimental studies were carried out by Gysbers et al. (2004) where the assumption was that the more spatial descriptors and instructions for the imagination would mean an increase in spatial presence for readers of a text. The results, however, indicated that the findings of the study only partially supported the assumptions and that the 'strength' of the SSM was not an important factor.

The Third Pole Proposal

As described above, presence is not only dependent upon and aided through the use of digital media, but in a non-mediated environment it has been described as 'the willing suspension of disbelief' (Coleridge, 1847); of being engaged by the representations of an imagined virtual world. Biocca (2003) argues that mediated presence is actually not as persuasive as non-technology mediated presence, and that the traditional concept of presence has not considered the notion of mental imagery space.

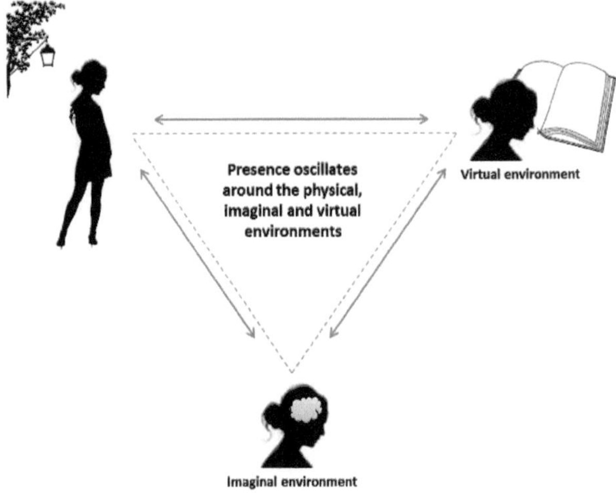

Figure 3.7: The oscillation of presence, in physical, virtual and imaginal environments. Adapted from Biocca (2003)

He argues that:

> The compelling sense of presence in virtual environments is unstable. At best it is fleeting. Like a voice interrupting a daydream in the imaginal environment, presence in the virtual environment can be interrupted by sensory cues from the physical environment and imperfections in the interface ... At one point in time, users can be said to feel as if they are physically present in only one of three places ...

> the physical environment, the virtual environment, or the imaginal environment. (Biocca, 2003, p. 21)

Traditional presence research postulates that presence occurs when the individual's experience oscillates between the physical world and the virtual world. However, Biocca (2003) argues that this does not account for the 'physical reality problem': the 'dream state problem' and the 'book problem', hence, the third axis of mental imagery was proposed, as illustrated in Figure 3.7.

According to Biocca (2003), a sense of presence can also occur when individuals have the impression that they have 'withdrawn' from physical space into a purely imagined space. This experience happens when the individual withdraws 'focal attention' from incoming sensory cues, and instead chooses to attend to 'internally generated mental imagery'. Furthermore, dreaming and daydreaming also reveal a space where one may feel present: the imaginary environment. "We can say that the user is present in the internally simulated, imaginal environment when the user:

- has withdrawn focal attention to incoming sensory cues,
- is attending to internally generated mental imagery,
- has diminished responsiveness to sensory cues from either the physical environment or the virtual environment" (Biocca, 2003, p. 6).

The mind is also capable of constructing convincing spatial environments in dreams as well as hallucinations and daydreaming (although the latter two are usually to a lesser degree). Within these environments, an individual consciously experiences a sense of moving through space, for example, interaction with objects or people or running from something that induces fear. Biocca argues: "Clearly in a dream state we are present in a spatial environment. But it is also clear that this environment has nothing to do with technology" (Biocca, 2003, p. 7), and that dreams use a form of cognitive 'simulator', which is the generator of mental imagery that makes use of cognitive resources used in perception (Biocca, 1997; Kosslyn, 1980). However, where typical states of presence depend on incoming sensory stimulation, the mental spatial simulation is mostly constructed from memory, although it is believed that in a dream state, cognitive simulation can be a response to environmental stimuli.

In the field of virtuality research, Biocca (2003) argues that research on presence has been focused toward a 'two pole model', where "the two pole model of presence posits that presence shifts back and forth from physical space to virtual space" (Biocca, 2003, p. 1), and although the two pole model may be useful in initial engineering research on remote operated telerobotics or telepresence, the model was erroneously generalised to all media and became a cognitive theory of presence (Biocca, 2003).

The original 'two pole model' of presence only considers virtual and physical spaces, but not imaginary spaces, argues Biocca, who proposes that the 'three pole model' allows for the role of mental imagery space. Furthermore, the two pole model fails to explain instances of high presence in media of low immersion, such as when one experiences a sense of presence while reading a novel (the book problem), as well as instances of low presence in physical reality, such as when one is present in a physical place but is relatively unaware of the place because they are mentally focused on something other than the immediate environment (the physical reality problem).

In proposing the three pole model, the 'mental imagery space' can be accounted for, and in addition, this 'third pole' allows spatial cues to be generated by mental imagery in addition to virtual or physical imagery. This mental imagery space, central to mental model development, explains why media of low immersion are capable of fostering a sense of presence in users.

In terms of mental imagery, Durand (1993) argues that "imaginary is the 'implicate' order through which all understanding necessarily passes, and even all explanations of individual or collective human behavior as well" (Durand, 1993, p. 17). Feshbach (1976) concurs: "an impressive amount of daily cognitive activity is fantasy in nature" (Feshback, 1976, p. 71). Regarding the overall role of imagination in cognitive activities, Kant (1933) argues that imagination is a necessary ingredient of the process of perception, distinguishing imagination (the power of synthesis) from both sensibility and understanding, and treated it as a separate faculty or 'subjective source of knowledge' (in Pendlebury, 1996). The representation of objects in the form of mental images is based upon the same cognitive mechanisms that are involved in the perception of those objects.

Mental imagery (occasionally referred to as visualisation, or 'seeing in the mind's eye') is an experience that resembles perceptual experience, but which occurs in the absence of the appropriate stimuli for the relevant perception (Finke, 1989; McKellar, 1957). These experiences are very often understood by the individual as echoes or reconstructions of actual perceptual experiences from their past; at other times they may seem to anticipate possible, often desired or feared, future experiences.

The imagining process is not usually conscious, and has been considered by researchers distinguishing between 'knowledge by acquaintance' versus 'knowledge by description', or experiential versus rational cognitive systems (Buck 1999; Epstein & Pacini, 1999). Mental images have an impact on an individual's information processing because, at times, it resembles regular perceptual processes and moreover, it is involved in memory storage and retrieval as well as in emotional experiences (Lang, 1979). Lang considered the image as an active response process and an emotional experience, rather than a stimulus in the head to which the individual responds.

In terms of the presence experience, 'induced imagery' (or imagination, or fantasy, or daydreaming) is combined with the actual mediated message in producing the final mental representation or response; imagination becomes part of the perceiving process itself (Palmgreen, 1971).

With presence research and its community, the discussion surrounding 'the book problem' encapsulates both the perceptual and conceptual perspectives in presence. The essence of the debate is the question whether less immersive media (such as text or comic art) are capable of inducing a presence experience. Biocca (2003) points out that the heart of the 'book problem' rests with the 'sensorimotor immersion assumption' which posits a direct correlation between the level of immersion of the medium and the level of presence experienced by the user:

> If sensorimotor immersion is the key variable that causes presence, then how do we explain the high levels of presence people report when

> reading books? Books are very low fidelity, non-iconic media and are extremely low on all sensormotor variables identified as causing presence: extent of sensory data, control of sensors, and ability to modify the environment. (Biocca, 2003, p. 4)

There are also a number of other theorists and researchers who recognise this incongruity (Banos, Botella, Guerrero, Liano, Alcaniz & Rey, 2005; Gysbers et. al., 2004; O'Neill & Benyon, 2003; Pinchbeck & Stevens, 2005; and Schubert & Crusius, 2002). Biocca (2003) notes that "most [theorists] see the illusion of presence as a product of all media" and that "few theorists argue that the experience of presence suddenly emerged with the arrival of virtual reality" (Biocca, 2003, p. 14); even those researching outside the domain of presence research argue that "low resolution media does not mean a low-resolution experience" (Phillips, 2000, p. 82).

Given this, Biocca proposes that shifts in presence (occurring when the individual experiences 'lapses' in presence during an experience; put simply, moving from physical space to virtual space) are likely to have 'predated media'. This space allows the addition and incorporation of spatial attention and mental imagery in the account of presence, and thus the 'third pole' is introduced. Spatial models, proposes Biocca, have similar properties to those of real and virtual sensorimotor spaces, and spatial presence, generated by mental imagery spaces, oscillates among three sources of spatial cues: real, virtual and self-generated (imaginal).

Schubert and Crusius (2002), in their formulation of the five theses discussion around the book problem, differ in their starting point of discussion on the issue, stating that presence is a cognitive construct created from immersive stimuli, rather than Biocca's claim that presence is a direct function of immersion. That is, the structure of a mental model determines whether or not an individual experiences a sense of presence. This means that in acknowledging this, the psychological phenomenon is the same in media, VR, film and text, and that the physical environment is "suppressed in favour of an alternative, medially presented and cognitively construed environment" (Schubert & Crusius, 2002, p. 262). Taking this into account, it is therefore important to acknowledge the role where cognition acts as a mediator between immersion and presence; the media should be regarded as the raw source of the mental model construction, but not directly responsible for the sense of presence. This means that the media allows the mental construction of an environment around the body, but it is the cognitive construct of the mind that permits presence to occur.

Schubert and Crusius (2002) offer a resolution to the paradox and propose a theory acknowledging a 'cognitive layer' to the experience of presence wherein all incoming perceptual stimuli do not give way directly to a sense of presence, but rather apply toward the construction of a mental model which may or may not induce presence depending on the level of detail. Presence is not a direct function of immersion; rather that immersion is dependent upon the source of a stimulus with which users create a mental model. It is the mental model that individuals create – which theoretically could be based on cues in virtual reality, television or books – that determines whether or not presence is felt (Schubert & Crusius, 2002). They posit that presence has been studied in various contexts under different names, including the 'diegetic effect' in film (Burch, 1979; Tan, 1996) and 'transportation' in narrative (Gerrig, 1993; Green & Brock, 2000).

An important argument proposed by Schubert and Crusius (2002) is that by differentiating between the construction of a spatial mental model and the attention devoted to this construction, some differences can be seen between modalities. Virtual reality, for instance, has a high potential for both construction of mental models, and therefore spatial presence, and for involvement (or attention), while literature has a high potential for involvement, but a weaker potential than both films and virtual reality for spatial presence (Schubert & Crusius, 2002).

Pinchbeck and Stevens (2005) support the conceptual view of 'the book problem', claiming that "the book problem should come as no surprise and rather than being an issue, should be taken as demonstrating that virtual environments and other media share the capacity to influence an organism's representation of its surroundings" (Pinchbeck & Stevens, 2005, p. 223). However, Waterworth and Waterworth (2003b) strongly oppose the conceptual view, asserting that presence is an external/perceptual phenomenon, and that the book problem is "a confusion between sense of presence and emotional and/or intellectual engagement in internal, imagined space" (Waterworth & Waterworth, 2003b, p. 1).

This poses an interesting debate, and the oppositional quality that the conceptual and external views and perspectives on the book problem have with respect to each other "reflects a fundamental difference in the understanding of what presence is and how it is constituted" (Jones, M.T., 2008, p. 30).

All in the Mind

The notion of being "lost in a book" is not a new discussion (Green & Brock, 2002; Green, Brock & Kaufman, 2004; Ryan, 2001) and the power of narrative has been documented – although not in great detail, according to Turner and Turner (2011), who argue that the book problem "is all in the mind" (Turner & Turner, 2011, p. 1): "Presence can be thought of as either the consequence of physically being-in-the-world or the product of technology which substitutes the real for the virtual. In both cases presence is dependent on sensory input" (Turner & Turner, 2011, p. 1).

Despite Biocca's (2002, 2003) discussion of 'the book problem' and the three-axis proposal, there are few in-depth studies and no agreed resolution of it, according to Turner and Turner (2011). Even with the inclusion of the third pole, Turner and Turner (2011) argue that there are still a number of unresolved questions. The first concerns the status of the other aspects of presence such as engagement, raising the question: which of the three axes contributes to the spatial presence experience? The second is: what exactly causes oscillation between the three poles (mental imagery space, virtual space, and physical space)? The final question is: why does imagination fade? In proposing these questions, Turner and Turner suggest: "Without a treatment of this movement, the three pole account remains a static account of spatial presence which flips from one state to another" (Turner & Turner, 2011, p. 2).

In response to the Schubert and Crusius (2002) five theses discussion around the book problem (where they suggest that presence is a cognitive construct created from immersive stimuli, rather than Biocca's claim that presence is a direct function of immersion), Turner and Turner (2011) argue that this raises an interesting question: how does one become 'lost' in a book? If, "to understand the book problem is to understand the nature and creation of the mental model" (Turner & Turner, 2011, p. 3), then the power of narrative should be examined.

Ryan's (2001) theories on narrative and presence (discussed above) illustrate that there is a cognitive link between narrative and presence; additionally, there is evidence that the reader creates a 'textual space' (Blackler, 2007) and is transported to story worlds by combining individual experiences and spatial cues, which together evoke mental imagery. However, there is still "sparse empirical work on being 'lost in a book'", and there are still "few firm data as to *why* (as contrasted to *how*) the written word is as powerful as full-spectrum virtual reality as a means of transportation" (Turner & Turner, 2011, p. 3). Taking this into consideration, they argue that "the phenomena of experiencing texts, on the one hand, and VR, films, and games on the other are fundamentally different" and the reasoning behind this is "because of the differences in the nature of the underlying cognitive and neural representations which mediate them" (Turner & Turner, 2011, p. 3). In order to illustrate this point, Clark's (1997a, 1997b) theories on cognitive science regarding 'weak' and 'strong' representations are illustrative.

A weak representation is that of an internal state that is "capable of bearing information about an external object only when that object is in close proximity" (Turner & Turner, 2011, p. 4). For example, an animal will be provided with quick feedback through what Clark describes as 'information and control systems' regarding a predator (the local external object) and thus will interact with that local object effectively, enabling a fight or flight mechanism in the animal. These 'systems' bear information about, and correlate with in a deliberate fashion, features of external objects. If the external object fades away and becomes absent or distant, the representation falls 'silent', but can be "stored off-line for future use or combined with other representations to form internal maps of the external world" (Turner & Turner, 2011, p. 4). In the case of media such as virtual reality, films and games, a person requires the external object (the source of the information – such as a game) to be within close proximity if presence is to be experienced.

On the other hand, a strong representation is considered to be the most genuine representative and the basis of cognitive processes; it is defined as "an information-bearing state that is serviceable even if its source object becomes distal or absent" (Turner & Turner, 2011, p. 4). Strong representation is "being 'de-coupled' from the immediate world and thus stand[s] for things when they are removed from us (or us from them)" (Turner & Turner, 2011, p. 6). Put simply, narrative or text (such as books) require the individual to use mental imagery based upon spatial descriptions to create a "textual space" (Blacker, 2007, in Turner & Turner, 2011, p. 4). The narrative itself is not the source of representation; rather, it is the mental imagery that the person 'constructs'.

Taking this into account, Turner and Turner (2011) consider that narrative is considered to create strong representation, but media such as virtual reality, films and games create weak representation: "the nature of the representation is independent of its medium or substance – what matters is how available it is/they are" (Turner & Turner, 2011, p. 5). Put simply, in considering the diverse set of research and literature on neuroscience, Turner and Turner (2011) argue that "the very same neural structures and processes are responsible for processing real and imagined stimuli" (Turner & Turner, 2011, p. 5) and that evidence suggests that when reading text, watching a movie, dreaming, or participating in a virtual environment, an individual experiences the same neural and cognitive processes.

In conclusion, Turner and Turner (2011) argue that the book, dreaming and real world 'problem' only developed due to the consequence of the "pervasive engineering paradigm in presence" (Turner & Turner, 2011, p. 6). In other words, the 'problem' with non-mechanical media – such as books, text, daydreaming – is that 'traditional presence research' argued that as there was no external device involved – such as a head mounted display – presence cannot be possible. Furthermore, they assert that whilst Biocca (2003) was correct that dreaming, real world, and books can create a sense of presence, he was incorrect in his proposition that imagery (the third pole) is the missing component in the original two pole model (where presence oscillates between the real and virtual). Rather, they concur with Pylyshyn (in Kosslyn et al., 2001) who proposes that mental images are not 'images' at all, but rather rely on mental descriptions that are no different than those that underlie language: "the pictorial aspects of imagery that are evident to conscious experience are entirely epiphenomenal" (Turner & Turner, 2011, p. 7). Therefore, there is no book problem, as the brain does not distinguish between the real and the imagined, "Nor is there a dream problem or a physical reality problem because all of these diverse sources of 'stimuli' are processed by the same parts of the brain. And the same parts of the brain give rise to very similar kinds of experiences – QED, no book problem" (Turner & Turner, 2011, p. 7).

Even with the ongoing debate regarding presence in non-technology mediated space (and including the book problem), Turner and Turner (2011), Biocca (2002, 2003), Schubert and Crusius (2002) have all been have instrumental in discussion regarding the possibility of experiencing presence in a seemingly very low immersion space, and thus the possibility of experiencing presence in an imagined space.

The Presence Divide

As this study focuses upon the conceptual processing of imagining in the design process, and, in turn, an individual's sense of being within an imagined space, it is deemed necessary to analyse the two separate conceptualisations of presence: the external/perceptual view of presence, and the internal/conceptual view of presence. Whilst this book is not focussed upon exploring the depths of immersion in narrative environments, what is important is the notion that presence may be elicited from media that are not digital and that presence may be experienced as an internal conceptual phenomenon as well as externally or perceptually.

A major distinction among presence definitions concerns whether presence is seen as an exclusively external/perceptual phenomenon or an internal/conceptual phenomenon. For example, Waterworth and Waterworth's (2001) definition of presence as "a conscious emphasis on direct perception of currently present stimuli rather than on conceptual processing" (Waterworth & Waterworth, 2001, p. 211) takes a clear stance on the side of external perception, whereas Biocca et al. (2003), in their definition of presence as "the phenomenal sense of 'being there' including automatic responses to spatial cues and the mental models of mediated spaces that create the illusion of place"

(Biocca et al., 2003, p. 459) takes the opposing internal/conceptual view.

To date, the seminal text in the field is M.T. Jones (2008, 2009), who has provided a comprehensive overview on both views of presence. This informs the discussion below.

Much debate has risen in presence research and there still remains a significant distinction to date among presence definitions concerning presence as an exclusively external/perceptual phenomenon, or an internal/conceptual phenomenon. For instance, one such example used by M.T. Jones (2008, 2009) is Waterworth and Waterworth (2001, 2003a, 2003b), who define presence as "a conscious emphasis on direct perception of currently present stimuli rather than on conceptual processing" (Waterworth & Waterworth, 2001, p. 211) taking a clear position on the side of external perception. Although Waterworth and Waterworth are not the only authors in adopting this perspective, to date they provide the most in-depth and thorough theoretical argument in favour of it. It is their research that shapes the next section discussing the external/perceptual perspective of presence.

The Biocca et al. (2003) definition of presence as "the phenomenal sense of 'being there' including automatic responses to spatial cues and the mental models of mediated spaces that create the illusion of place" (Biocca et al., 2003, p. 459) takes an opposing view to Waterworth and Waterworth, stating that presence is an internal or conceptual process and is not external perception. The primary source of contention between these opposing views is 'the book problem', which was described in the previous section. Biocca (2002, 2003) and M.T. Jones (2008, 2009) still remain leading authors in the field of presence research who argue in favour of the internal/conceptual model of presence. To date, M. T. Jones' (2008, 2009) seminal research on the review and framework of the internal/conceptual and the external/perceptual divide is the most comprehensive surrounding the discussion.

In proposing that presence may be more conceptual or internally influenced, Weibel, Wissmath & Mast (2011), in their study on the relationship between mental imagery and enjoyment, found that mental imagery ability plays an important role in presence and enjoyment, whereby their particular role strongly depends on the medium. They conclude that individual mental imagery abilities are an interesting source of information when studying media effects and presence. They suggest that future presence research should consider imagery as a potential confounding variable in the presence experience.

On the other hand, Wirth et al. (2007) propose that "spatial imagination becomes more relevant if the mediated representation of the space is less intuitive and more fragmented (e.g., when reading textual descriptions)" (Wirth et al., 2007, p. 502) and this aligns with Wallach, Safir and Samana (2010), who argue that imagination does not have a significant impact on presence in sensory-rich environments. Jacobson (2002), however, suggests that imagery abilities may play a role when sensorial cues are less available, and they can, to some extent, stand in for the missing perceptual information (Jacobson, 2002), and this is the case with reading books. Despite the absence of any immediate perceptual stimulation, the reader is able to experience a high degree of immersion (Green, Brock & Kaufman, 2004).

According to simulation theory (Oatley, 1994), the story becomes convincing and enjoyable because the readers create the plot in their mind based on their own experiences. Mental simulations run on in the mind while reading, and imagined worlds are created, which allow for immersion in the narrative of a book (Oatley, 1994). Accordingly, Green, Brock and Kaufman (2004) assume that individuals with good visual imagery abilities are able to immerse themselves more in books than in visual media, suggesting that presence is more internal than external processing.

Perceptual Presence

In describing the philosophical roots of the external or perceptual view of presence, Biocca (1997) observes: "Many immersive virtual reality designers tend to be implicitly or explicitly Gibsonian" (Biocca, 1997, p. 2). What Biocca refers to here is the sense that many start with the assumption that no pre-existing knowledge of the world is necessary in order to make sense of it, as presence within an environment is constituted through direct perception. To quote Gibson: "The young child does not need to have ideas of space in order to see the surfaces around him" (Gibson, 1979, p. 304). As such, this assertion follows Gibson's argument that human visual perception evolved based upon the extraction of variants from the flux of the environment. In other words, thinking that visual perception requires predetermined, static concepts is in fact incorrect, because our system of visual perception evolved within perpetual flux; we perceive objects from mobile senses and objects, such as our moving eyes, our moving head on a moving body, and we perceive aspects of a mobile environment. According to this theory, only the extraction of the unchanging aspects of the environment is necessary to comprehend and apprehend the surrounding world (Biocca, 1997).

Waterworth and Waterworth (2001, 2003a, 2003b); Waterworth et al. (2001); Riva and Waterworth (2003); and Slater et al (2003) have all presented significant theoretical and empirical evidence that supports a view of presence that is based exclusively on external or perceptual phenomena. As mentioned above, the following debate on the external or perceptual perspective is based principally on the work of Waterworth and Waterworth (2001, 2003a, 2003b), not because they are alone in adopting this perspective, but because they have articulated the most thorough theoretical argument in favour of it. It should be noted that this perspective is widely prevalent within the presence community and is found in the predominantly engineering, psychology and communication foci in journals such as *Presence: Teleoperators and Virtual Environments* (Durlach & Slater, 1992), *Cyberpsychology and Behaviour*, and *Journal of Computer-Mediated Communication.*

The argument for justifying the external/perceptual perspective is through examining presence as part of the human evolutionary history (Waterworth & Waterworth 2003a, 2003b). These authors draw a distinction between 'core consciousness' (refer Damasio, 1999) and 'extended consciousness', suggesting that core consciousness is what we have in common with all conscious creatures, enabling understanding of our immediate concrete environment; extended consciousness is a capability distinctive to humans, allowing us to imagine possibilities and consequences as well as the ability to plan for the future. What is essential to the capacity of extended consciousness is the ability to distinguish between the domains of core consciousness (that is, perception of the immediate physical environment) and extended consciousness (that is, imagination), since to confuse the two would be dangerous and affect humans in terms of the ability to evolve:

> Viable organisms must be able to tell the difference between an imagined future situation and the actual, present, external situation. Confusions of the two indicate serious psychological problems, problems which, until recent times, would have prevented survival and the passing on of this condition. Simply put, if we react as if the external world is only imaginary we will not survive long (think of this the next time you cross a busy street). And if we think that what we are merely imagining is actually happening, we may omit to carry out basic activities on which our survival depends. We are suggesting that presence is the feeling that evolution has given us to make this vital distinction. (Waterworth & Waterworth, 2003b, p. 4)

These authors maintain that it is the sense of presence that distinguishes the difference between extended consciousness and core consciousness, with presence being within the domain of the latter.

Presence and Absence

Earlier research by Waterworth and Waterworth (2001) supports the view that presence is strictly perceptual and they propose that the term 'absence' should be used for activities such as imagining and thinking. They define that absence is "characterized as a psychological focus on conceptual processing, and presence as a psychological focus on direct perceptual processing (of things that are present in the current environment, whether real or virtual)" (Waterworth & Waterworth, 2001, p. 203). Further to this, they put forward a metaphor of the mind being a "two-room apartment" (Waterworth & Waterworth, 2001, p. 205), setting up presence and absence in an oppositionally and mutually exclusive arrangement by using the imagery of a sectional view of two adjacent rooms with a hanging light situated at the top of the doorframe between the rooms. The room on the left represents concrete processing (presence) where the body and world are represented as *being*, whilst the room on the right represents abstract processing (absence), where concepts and memory are represented as *doing*. The light areas represent conscious and the areas in darkness, unconscious.

Imagining if the lamp (which symbolises conscious thought) could be shone into one room or the other (but not simultaneously), suggests that consciousness is a zero sum game and that the concrete and abstract realms of thought compete as thus: "Although dreams, daydreams, images, fantasies, and other mental experiences may be extremely vivid, they are in competition with the experiences underlying presence in real or virtual worlds. Put simply, you cannot feel present in a virtual world, or in the real one, while also being lost in thoughts, dreams, or fantasies" (Waterworth & Waterworth, 2001, pp. 206-207). Therefore, following this line of reasoning, if one is conscious of the immediate world outside of the body, one is considered present, and if one is mostly conscious of one's own thoughts, they are absent.

In presenting the three dimensions of presence (focus, locus and sensus) (Waterworth & Waterworth, 2001), it is apparent that only two (sensus and focus) either enhance or detract from the experience of presence. Locus only distinguishes whether attention is directed toward the physical (real) or virtual (artificial) external worlds. Both are capable of producing a sense of presence. Sensus is the level of consciousness ranging from unconsciousness or asleep (absent), to fully alert (present); and focus is the variable that distinguishes concrete processing (presence) from abstract processing (absence).

In terms of testing external/perceptual presence, there have been numerous validation tests, most notably, Waterworth and Waterworth (2001) and Waterworth et al. (2001). Others include Banos et al. (2005) who measured the telepresence levels of participants as they explored a virtual park or an imagined park. These examples back up the enduring criticism levelled against the internal/conceptual view of presence, that the term 'presence' is becoming more conflated with the more general and already well-researched area of conscious attention. In fact, Waterworth and Waterworth (2003b) note that "in trying to solve the so-called book and dream-state problems the baby of presence has been thrown out with the bathwater of conscious attention; there is nothing left for the concept of presence to do" (Waterworth & Waterworth, 2003b, p. 1).

Criticisms of the external/perceptual view of presence

The external /perceptual view of presence endorsed by Waterworth and Waterworth and others sets up a logical criterion for discriminating between the experience of presence and 'absence'; it also leads to an understanding of the factors and contexts that enhance and detract from the experience of presence. Conversely, it contains a number of implicit assumptions that efficiently steer clear of the more complex examination of cognition that may be prompted if made more explicit. This 'logical' criterion sits more neatly with the 'hard sciences' (such as engineering), which depend on more quantifiable phenomena, rather than the 'soft sciences' (such as design and the humanities), which are more open to phenomena that are qualitative based.

In critiquing Biocca's 'three-pole model of presence' (2003), Waterworth and Waterworth (2003b) tacitly take for granted that the only role conscious attention has to play is as a necessary precondition of the presence experience. In terms of the sensus dimension, conscious attention serves solely to permit the prerequisite perceptual resources to be allocated to the immediate surrounding environment. This hidden assumption is that conscious attention occurs independently of cognition, which, according to this perspective, is not a determinant of presence, but rather is of the opposing condition referred to as 'absence'. The key issue here is that it is difficult to define conscious attention without in some way referring to cognition.

An example of this is Hu, Janse and Kong (2005), who provide a typical definition of attention as "a cognitive process of selectively concentrating on one thing while deliberately ignoring other things" (Hu, Janse & Kong, 2005, p. 4). Wu (2011) further defines conscious attention:

> Perceptual attention is essential to both thought and agency, for there is arguably no demonstrative thought or bodily action without it. Psychologists and philosophers since William James have taken attention to be a ubiquitous and distinctive form of consciousness, one that leaves a characteristic mark on perceptual experience. As a process of selecting specific

> perceptual inputs, attention influences the way things perceptually appear. It may then seem that it is a specific feature of perceptual representation that constitutes what it is like to consciously attend to an object. In fact conscious attention is more complicated...the phenomenology of conscious attention to what is perceived involves not just a way of perceptually locking on to a specific object. It necessarily involves a way of cognitively locking on to it as well. (Wu, 2011, p. 1)

Therefore, deliberate ignoring and selective concentration are activities that require cognitive effort. The matter of conscious attention stems from a larger problem with the external/perceptual position: the denial of the role that cognition plays in perception. Of course, it is possible for the presence experience to rely upon cognitive processes and conscious attention without being confused by them. However, as Ryan (2001) notes: "It [the mimetic concept of immersion/presence] applies to novels, movies, drama, representational paintings, and those computer games that cast the user in the role of a character in a story, but not to philosophical works, music, and purely abstract games such as bridge, chess and Tetris, no matter how absorbing these experiences can be" (Ryan, 2001, pp. 14-15). In the first set of examples, content tends to be narrative and to portray natural environments, whereas, in the second set of examples it is more abstract. The important aspect is that all involve both cognition *as well as* perception. The distinction that determines the presence experience should not be made between cognition and perception, but rather, among the *contents* of cognition.

In the discussion that denies the role that cognition plays in perception, Waterworth and Waterworth (2003a) maintain that a significant criterion for the sense of presence is the experience of an external, sharable world that yields the same perceptions among different individuals. This argument itself poses questions raised by phenomenologists and psychologists alike, but whilst attempting to make clear the distinction between a virtual or real world and an imagined world, they note that "The virtual world is the same for everyone who acts in it, just as the real world is," yet this conflicts with their own admission qualifying that "our experiences and reactions differ" (Waterworth & Waterworth 2003, p. 8).

Another issue with the external/perceptual view of presence is the use of the term 'absence' which describes non-presence in conceptual processing, illustrated in Figure 3.7 above. Although labelling this conceptual space as 'absence' makes a greater case for the distinction between 'external' and 'internal' perspectives, as well as the presence concept, there is a risk that the deeper underlying phenomenon may be overlooked for the sake of simplicity. Within the area of memory processing and dreams, where previously experienced physical locations can be recalled, the argument for the ability to invoke presence can be made. In other words, the person, through recalling the space, or experiencing the space in the mind's eye, can actually feel as though they are there. Many individuals have had the experience of dreaming about something, and, at the time, it seemed completely real.

Rassin et al. (2001) describe their study based upon dreaming where a considerable amount of undergraduate students reported the experience of false memories based upon the inability to discriminate between perceptually realistic dreams and waking reality. If these phenomena are considered to be absence, based upon their conceptual nature, it would seem that an understanding of presence as a subjective psychological state is no longer salient as it is now determined based on subjective location of the body, regardless of where the mind is focused.

When examining the literature, it becomes apparent that the notion of absence may have been derived from the term 'absent minded'; when an individual undertakes difficult abstract reasoning, time passes quickly, and little attention is paid to one's own body. It is a state where we do not feel present in the world (Waterworth & Waterworth, 2001); whereas it is argued that when our conscious processing load is light, time passes quickly and one feels highly present in the surrounding environment. However, this raises the question of the dichotomy of 'being absent' and 'being present'. Even whilst a person is working though deeply abstract processing and may feel absent from the surrounding world, it is still possible that they may feel present in a world that is 'not of this world'.

The external/perceptual paradox

Criticism of the external/perceptual view concerns how the studies related to it were interpreted. Initially, when examining the experimental evidence described above, it appears to favour the external/perceptual view of presence, yet it also serves to support the conceptual model for several reasons.

First, each experiment indicates that media which requires internal/conceptual processing such as imagery instructions (Banos et al., 2005) and abstract content (Waterworth et al., 2001) , whilst providing a lower level of experienced presence still elicits some form of presence, and this is presumably proportionate with individual imaginative capabilities. If the proposition is true that media requiring conceptual processing did not produce presence, but instead produced 'absence' because of the conceptual component involved, then the levels of reported presence should not be recorded as being lower than immersive media; it should be recorded as zero. By the definition proposed by Waterworth et al. (2001), one cannot read or engage in imaginative activities without entering the space that Waterworth and Waterworth (2001) designated as 'absence'.

Pinchbeck and Stevens (2005) assert that "Simply stating that reported presence from media with low immersive capabilities is not presence but something fundamentally different, if indistinguishable when using existing measures, is an unacceptable stance" and that "To view presence as either a unitary, or a uniquely perceptual construct is untenable. Instead, it appears more likely that presence is an emergent property of a combination of cognitive and perceptual processes and stimuli" (Pinchbeck & Stevens, 2005, p. 222).

Second, the argument that the more immersive form of media, the more it produces an intense presence experience does not mean that cognition has no part in the process. If, as Waterworth and Waterworth (2001) suggest, the incoming stimuli (regardless of the level of immersive quality) serve only as the raw material out of which a mental model is constructed (Biocca, 2003; Schubert & Crusius, 2002; Schubert, Friedman & Regenbrecht, 2001) then an initially more complete environment would be naturally easier to process cognitively, thus reducing conceptual tasks and providing a more intense presence experience.

Nuñez and Blake (2003a) report on experiments that shared several similarities in examining the potential relationship between presentation quality and the sense of presence. The experiment was conducted as part of a larger study into the cognitive processing of participants in virtual environments. A direct comparison of presence scores was collected from three separate groups of users each visiting the same virtual environment but on a different display system. The first group were on a high quality graphical system, the second on a low quality graphical system, and the third group on a text-based system. The purpose of the experiment was to investigate how text displays would compare to graphical displays with regard to presence elicited. The measurement of presence levels were by means of the Slater et al. (2003) presence scale, and the Presence Questionnaire of Witmer and Singer (2005). The results confirmed that graphics-based virtual environments produce statistically higher levels of presence than text-based systems, but the actual difference between both produced was quite small. The authors proposed that the difference can be regarded, for many practical applications, as negligible, and "the designer of text-based virtual environments can expect the presence on those systems to be lower than in graphics-based systems, but by less than 20%. Although lower presence implies a reduction in benefits such as task performance enhancement and effective navigation, the fact that the difference is small in turn implies that the reduction of these benefits will also be small" (Nuñez & Blake, 2003b, p. 3).

Quantitative methods for measuring the presence experience (such as the Presence Questionnaire) have been accepted and embraced within the presence research community for a reasonable amount of time and many experiments have been carried out utilising them. Whilst using quantitative methods does aim for a sense of consistency, on the other hand, qualitative methods provide rich and detailed data that reflect the experiential and phenomenological aspects of the presence experience. However, qualitative methods are still underutilised in presence studies to date (2012).

It can be argued that the studies carried out by researchers such as Waterworth and Waterworth (2001) may not, in fact, measure the *experience of presence* itself, but rather, they measure the *effects of presence.* Thus, it provides feedback that may in fact illustrate that defining presence as 'present' and imagining as 'absent' are somewhat narrow conceptualisations and do not explore either definition with significant complexity. Presence in itself is an experience that has both tangible as well as intangible effects. Having these qualities, this phenomenon requires the understanding of not only the *quantity* of the experience, but also the *quality* of the experience. The use of Likert scales (which are commonly in presence research) to measure an experience may not be the best method to understand such a complex phenomenon as presence. On the one hand, the benefit of these types of scales is that questions used are usually easy to understand and so lead to consistent answers. Conversely, a disadvantage is that only a few options are offered, with which respondents may not fully agree; and issues can occur where people may become influenced by the way they have answered previous questions. Likert (1932) gives an example such as if a respondent has agreed several times in a row, they may continue to agree. They may also deliberately break the pattern, disagreeing with a statement with which they might otherwise have agreed. This patterning can be broken up by asking reversal questions, where the sense of the question is reversed – thus in the example above, a reversal might be 'I do not like going to Chinese restaurants'. Sometimes the 'do not' is emphasised in order to ensure the respondent notices the difference in questions, although this can cause bias and hence caution should be applied when using it in surveys.

Another issue is that there is much debate about how many choices should be offered in a questionnaire. That is, confusion is observed in having an odd number of responses. Midpoint neutral statement of "neither agree nor disagree" is confused with "don't know" or "not available" (Raaijmakers, Van Hoof, Hart, Verbogt & Vollebergh, 2000). An odd number of choices allows participants to 'sit on the fence', whereas an even number forces people to make a choice, whether this reflects their true position or not.

The use of Likert scales is also argued to contravene one of the important principles of formulating an instrument: clarity and conciseness. That each Likert scale item measures more than one dimension at a time is considered to increase cognitive complexity, thus elevating measurement error (Hodge & Gillespie, 2003); each item of the scale measures both directions (agree/disagree) and strengths (strongly, or not so). This can also make the most extreme positions (strongly agree/disagree) under-reported (Albaum, 1997).

A related criticism of the external/perceptual model is the method of determining if a sense of presence has been achieved. Whilst there are several neurophysiological phenomena studies testing brain activity (EEG) and its relation to presence experienced (Schlögl, Neuper & Pfurtscheller, 2002), in terms of outward physical response, Waterworth and Waterworth (2001) adopt Nunez and Blake's (2001) reference of 'behavioural presence' or the 'postural or movement approach' (Nunez & Blake, 2001, p. 115). This, they state, distinguishes presence from mere conscious attention: "The reader of a novel may become deeply engrossed in the lives of the characters and the action that is described, but they are unlikely to move their bodies unconsciously to avoid a hazard that is only described in text" (Waterworth & Waterworth, 2001, pp. 204-205). However, while there appears to be little disagreement within the presence community as to the prerequisite that a level of physical response is required whilst the individual is experiencing a sense of presence, within this debate, the extent to which this is the case is yet to be determined.

The next section offers a discussion of the internal/conceptual view of presence, the cognitive processes involved in producing a mental model to which the individual responds, the suspension of disbelief and facilitating mental simulation, providing an insight into the processes by which presence occurs in 'low-immersive' media.

Conceptual Presence

In discussion regarding presence and its various perspectives, it is important to note their emergence within the field of telerobotics engineering/'hard sciences'. Since its inception, presence research has developed substantially to recognise the capacity of other fields of research and other media forms to provoke the same essential experience, although the debate regarding 'non-technological media' eliciting a sense of presence still persists.

The mental model

According to Schubert, Friedmann and Regenbrecht (2001), presence is an experience which is derived from interaction with a mental model of the surrounding environment. This aligns with similar theories that have been applied to the process of reading such as Oatley (1999), who posits: "human mental life depends strongly on constructive abilities. What human minds do generally is to make models that parallel the workings of the world" (Oatley, 1999, p. 105). However, the most significant point in terms of presence is the individual's interpretation of their mental model, for it is within this internal conceptual act that a sense of presence is felt (Schubert et al., 2001); what is most important about this is to recognise that the mental model does not only apply to technologically mediated involvement. In the *Five Theses* (Schubert & Crusius, 2002), when referring to the "cognitive representations as another theoretical layer", they mean that all perceptual cues (whether originating in the physical, virtual or imaginary environments) serve to construct an internal representation that we react to; that is, they are all filtered through cognitive representations. A sense of presence may then result from a distal attribution of that internal model (Biocca, 2003).

Distal attribution is defined by Loomis (1992) as a phenomenon where "most of our perceptual experience, though originating with stimulation of our sense organs, is referred to external space beyond the limits of the sensory organs" (Loomis, 1992, p. 113). In relation to the theory of distal attribution is 'embodied cognition' (Beer, 1995; Clark, 1998; Greeno & Moore, 1993; Thelen & Smith, 1997; Wertsch, 1998) where the environment is regarded as a part of the cognitive system. Furthermore, "the forces that drive cognitive activity do not reside solely inside the head of the individual, but instead are distributed across the individual and the situation as they interact" (Wilson, 2002). Simply put, the model that we construct from within becomes mapped onto, or attributed to, the external environment.

Schubert et al. (2001) argue that the experience of presence in text, film and virtual reality originate from the same cognitive process; when in the process of the presence experience, the mind both *constructs* objects and entities, as well as *suppresses* irrelevant information.

In particular, the authors of the MEC Model of Spatial Presence (Vorderer et al., 2003) stress the importance of the role of the mental model in the presence experience. As discussed previously, a spatial simulation model is formed based upon two primary components of information; the 'top-down component, which relies upon the implementation of pre-existing knowledge to construct the mental model, and the 'bottom-up' component, which constructs the mental model based upon descriptive information.

The core elements of both the suppression/construction model and the top-down/bottom-up model both form the basis for the discussion in the following sections that describe how the mental model functions to form a sense of presence. Mental simulation, which has been described as an embodied, grounded, and situated cognitive activity, plays an essential role in bottom-up processing and is the functional act that produces the structure of the mental model. Biocca (2003) sees mental activities such as dreams, daydreams and hallucinations as evidence that the mind is able to produce "compelling spatial environments" (Biocca, 2003, p. 17) and Schubert et al. (2001) compare mental model construction to language comprehension and memory.

Related to mental simulation is 'grounded' or 'embodied cognition', which is where the human cognitive activity is grounded in sensory-motor processes and situated in specific contexts and situations. Therefore, in this view, concepts consist of the reactivation of the same neural pattern that is present when we perceive and/or interact with the objects they refer to. Both mental simulation and grounded/embodied cognition are two such concepts that may be instrumental in constructing simulations of the present and future as called for by acts of imagination.

Suspension of disbelief

A concept that has been discussed at length in the literature on fiction, film and presence is the suspension of disbelief (Coleridge, 1847), an action that appears integral to the act of mental simulation. As the activity of engaging in narrative does require engagement and cognitive effort, willingness and motivation on the part of the reader (Gerrig, 1993; Gysbers et al., 2004; Ryan, 2001) that initial step towards receptivity to the narrative requires explanation.

Biocca (2003), in defining the experience of presence in the imaginal environment, describes it in terms that are rather similar to the concept of the suspension of disbelief, noting that diminished attention and responsiveness to sensory cues in the immediate environment (versus the virtual environment) is a prerequisite to presence. In a similar capacity, suppressing the surrounding physical environment is a task that is essential to involvement (Schubert, 2003; Schubert et al., 2001), memory, and language comprehension

(Glenberg, 1997). Suppression is therefore accomplished through the suspension of disbelief on the part of the reader. Further to Glenberg's (1997) example involving language comprehension, he points out that the suspension of disbelief (although not specifically referred to as such) is accomplished through the suppression of the physical environment and the structure of language itself. Therefore, in order to 'run' a mental simulation, the physical world must be suppressed, or 'left behind', and the physical symbols that signify the mediated world must 'fall away' to reveal the connotations they were crafted to produce.

However, according to the proponents of the external/perceptual view of presence, the suspension of disbelief alone is deemed insufficient to invoke telepresence:

> The root of the problem with many existing models of presence is perhaps confusion between presence and suspension of disbelief. Suspension of disbelief is the result of conceptual processing which then leads to a secondary sense of involvement – as when we read a gripping novel in which we become engrossed. On the other hand, we see presence as that which arises in situations where no belief suspension is needed, because the display is immediately perceptually engaging. (Waterworth & Waterworth, 2001, p. 204)

One potential problem with this perspective is that it neglects the notion that the suspension of disbelief is required for highly immersive virtual experiences. For example, it could be argued that the sensation of the weight of a head-mounted display must require the suspension of disbelief arising from the weight or pressure on the head in order to experience a sense of telepresence. Put another way, devices that are used to create a display can encroach on the senses in a way that does not correspond to the content of the display, serving as constant reminders of the mediated nature of the experience. In almost every technology intended to foster telepresence, there are numerous examples of this problem, and to date, there are very few ways in which to completely rid of certain devices that are created for the mediated telepresence experience.

In some contexts, such as text-based virtual environments, a functional transfer of knowledge between 'real' and 'virtual' about learning experiences can be reached by not providing all the details and information typical of a high fidelity reconstruction, but leaving users in a particular 'suspended position': they only receive a few details about the virtual context and starting from this incomplete level of information, they can 'complete' the lack of definition in virtual environments constructing the scenario. This is the model of construction detailed above (Schubert et al., 2001). In this process of knowledge construction, imagination plays a major role: "although detailed description is critical in fostering a sense of presence in imagined and imaginary worlds, novelists, literary theorists, and other scholars agree that details must be selective and that depth of detail should not undermine the "glory of imagined description" (Schubert et al., 2001, p. 655).

Several other researchers agree with this notion of construction, underlining the role of imagination in experiencing presence (Biocca & Levy, 1995; Ryan, 1991), with Reed (1991) stating that "readers want only a few external details to confine their imaginations; they will take what the writer has given and supply the rest" (Reed, 1991, p. 5) The imagination is seen as filling in the gaps of information facilitating and reinforcing the involvement and engagement also in textual materials (Jacobson, 2001).

According to Schiano and White's (1998) opinion, some individuals prefer text-based environments (not audio and video) in order to represent themselves, thus giving the individual a greater sense of agency; imagination is critical for the sense of presence. For this reason, Gerrig and Pillow (1998) change the definition of presence from 'willing suspension of disbelief' into 'willing construction of disbelief' where people believe and support what they read or experience.

Conversely, the notion of 'anomalous response', proposed by Gerrig (1993), does not rely on suspension of disbelief to explain mental simulation as it occurs in the experience of narratives. He suggests that there is something deeper than intentional 'ignorance' at work:

> I reject simple 'toggle' theories of fiction which have suggested that readers perform some mental act called 'the willing suspension of

> disbelief' that eviscerates the effects of fiction. Whatever the effects of narrative worlds may be, they will arise because of strategic actions that experiences do and do not perform with respect to those worlds. (Gerrig, 1993, p. 17)

Replacing the notion of the suspension of disbelief, Gerrig suggests that anomalous suspense explains how the narrative world comes to take precedence over the individual's immediate surroundings. The phenomenon of anomalous suspense describes reader suspense under conditions in which their real-world knowledge should prevent the sensation of suspense.

In order to test this concept and demonstrate its salience, Gerrig set up two conditions in which experimental participants must read a story and respond to questions. In the first condition, the story is written in such a way as to inspire suspense in the reader while, in the other condition, the story is not written to inspire suspense. The results showed that, on average, participants in the suspense condition took significantly longer to determine the truth of factual outcomes (that they had knowledge of beforehand) than participants in the non-suspense condition (Gerrig, 1993). Gerrig attributes this finding to the reader's tendency to consider the potential conclusions insinuated by the text. In other words, when a set of hypothetical circumstances or conditions are presented that provide a congruent internal framework, alternative scenarios are entertained in spite of real-world awareness. Gerrig further argues that this occurs because of an "expectation of uniqueness" (Gerrig, 1993, p. 170) that we experience while progressing through the event structure of a narrative. Such an expectation, he suggests, derives from an "optimization of cognitive resources" (Gerrig, 1993, p. 170) that evolved from our interactions with physical reality, which never quite repeats itself the way our own manufactured narratives do.

M.T. Jones (2008, 2009) argues that is a reasonable conclusion considering that throughout the majority of our development as a species the precise repetition of an event (or even a story that is told orally) has rarely, if ever, been encountered. With the invention of recorded narrative, details are held standard even although on some primitive level, we never expect to encounter the identical set of details when revisiting the same narrative despite the fact that, logically, we should. In this sense, mental simulation should be viewed as an active process which does not run a standard program of simulation the way, for instance, a sound file played in a stereo might simulate the sound of an orchestra; it is an active process of creation and recreation that takes on its character at the moment of inception.

Similar to this notion of improvised creativity are children's imaginative games. Harris, Brown, Marriott, Whittal, and Harmer (1991) tested the notion of creativity in an experiment: using two empty boxes, they asked children to pretend that a particular box contained either a monster or a rabbit. The findings showed that children interacted with that box differently (approaching and touching it), once the experimenter had left, even though the children had verified and acknowledged that the suspect box was empty. Gerrig (1993) interprets this in terms of his anomalous suspense theory: "When subjects become immersed in a narrative world in which the box is occupied, they might find that because attention is so thoroughly captured, knowledge of the box's real world emptiness is momentarily crowded out of consciousness" (Gerrig, 1993, p. 194).

Facilitating mental simulation

Biocca (2003) reminds us that Munsterberg, the first psychologist to study media in 1916, hinted at an issue which has not always been not taken into account when examining presence and its effects: media obey laws of the mind. Presence is a user experience and it is not intrinsically bound to any specific type of technology, but is rather a product of the mind (IJsselsteijn & Riva, 2003). Biocca (2003) also reminds us of the celebrated phrase of Bricken, from the 1990 SIGGRAPH (Special Interest Group on GRAPHics and Interactive Techniques) conference: "Psychology is the physics of virtual reality". According to Biocca, this sentence implies that, like physics, psychology holds a key to our understanding of reality. Therefore, VR "has less to do with simulating physical reality per se; rather it simulates how the mind 'perceives' physical reality" (Biocca, 2003, p. 18).

In moving the focus from *user characteristics* to *media/medium characteristics*, it is important to discuss some features of low-immersion media (such as text) that may be manipulated to ease the process of mental simulation; this in turn creates a stronger mental model that is more capable of inducing presence. Media requiring much mental imagery and conceptual processing often use strategies to transcend the medium; Joseph Conrad reinforces this through his often cited quote: "My task which I am trying to achieve is, by the power of the written word, to make you hear, to make you feel – it is, before all, to make you see" (Conrad, 1951, p. 26).

Coleridge (1847), in his *Biographia Literaria*, in the context of the creation and reading of poetry, also attempted to achieve the same goal through the 'suspension of disbelief'. Poetry and fiction involving the supernatural had lost its popularity to a large extent in the eighteenth century, in part due to the declining belief in witches and other supernatural agents among the educated classes who embraced the rational approach to the world offered by the new science. Coleridge's concept of 'willing suspension of disbelief' explained how a modern, enlightened audience might continue to enjoy such types of story, with the author recalling:

> It was agreed, that my endeavors should be directed to persons and characters supernatural, or at least romantic, yet so as to transfer from our inward nature a human interest and a semblance of truth sufficient to procure for these shadows of imagination that willing suspension of disbelief for the moment, which constitutes poetic faith. Mr. Wordsworth on the other hand was to propose to himself as his object, to give the charm of novelty to things of every day, and to excite a feeling analogous to the supernatural, by awakening the mind's attention from the lethargy of custom, and directing it to the loveliness and the wonders of the world before us. (Coleridge, 1847, Chapter 24)

Both examples reveal the author's intention to use language in such as way as to access the perceptual senses of the reader through the written word.

Birkerts (1994) and Ryan (2001) believe that the connotative and denotative capacity of words and symbol allow abstract codes to be processed in such a way as to draw attention to what they signify rather than their own particular characteristics as signifiers. Birkerts goes further to explain that "reading is a conversation, a turning of codes into contents" (Birkerts, 1994, p. 97) and more emphatically, Glenberg (1997) notes that "we understand language by creating embodied conceptualisations of situations the language is describing" (Glenberg, 1997, p. 12). Responsible for the experience of presence is the formation of these embodied conceptualisations which are at the heart of the process.

Discussion and Conclusion

Although research on presence has gained new momentum and direction within the past five years, researchers are recognising that cognitive theories of spatial presence are needed, and rather than focussing on the question *what is presence*, it is now *how does presence occur* that has become the focus. Therefore, rather than emphasising aspects of the *measurement* of presence, attempting to create theories about the *process* of the presence experience offers a significant contribution to the field. However, while the current debate is on the cognitive and experiential aspects of presence, to date there still remains a significant absence of discussion linking the cognitive aspects of other disciplines such as imagining in the design process.

Presence research and its connection with agency provide enormous possibilities for the exploration of environments allowing different opportunities supporting users and their intentions. However, within this, there are still opportunities to scrutinise different forms of presence, encompassing perception, neurological processing, cognition, interaction and emotions, affecting people and culture, especially in the disciplines of architecture and interior design, as well as in the process of designing. This raises an important question: how can these various characteristics of presence allow the design and development of essential elements that capture the salient aspects of human presence and interaction within spatial design?

The two separate conceptualisations of presence – the external/perceptual view and the internal/conceptual view of presence – are essentially opposing views yet both raise valid arguments perhaps suggesting that presence should not be viewed as only external/perceptual or internal/conceptual. 'The book problem' (Biocca, 2002, 2003; Schubert & Crusius, 2002), where the possibility of experiencing presence in a seemingly very low immersion form of media is a strong foundation for the proposition that an individual may experience presence in a situation of imagining during the design process.

Where the original 'two pole model' of presence considered virtual and physical spaces, but not imaginary spaces, Biocca's (2003) 'three pole model' allows for the role of mental imagery space. However, where this model takes the conceptualisation of presence to another level, to date, there is still room for further studies which include imagined spaces that architects and designers utilise in their practice of designing on a daily basis, either to solve problems, to 'understand the space' (cognitively), or to 'experience' other non-tangible possibilities the space may contain. In proposing the three pole model, the 'mental imagery space' can be accounted for; it allows spatial cues which contribute to the mental model, thus in turn, facilitating presence.

Setting up a logical criterion for discriminating between the experience of presence and 'absence', the external /perceptual view of presence endorsed by Waterworth and Waterworth (2003, 2003b) and others leads to an understanding of the factors and contexts that enhance and detract from the experience of presence. On the other hand, this view also contains a number of implicit assumptions that efficiently steer clear of the more complex examination of cognition. Although the external/perceptual view 'fits' more neatly with the disciplines such as engineering, which depend on more quantifiable phenomena, it does not necessarily 'fit' as well with disciplines such as the humanities, and architecture and design, which are more open to qualitative based phenomenon.

The term 'absence' used in the external/perceptual view depicts non-presence in conceptual processing; describing this conceptual space as 'absence' makes a greater case for the distinction between 'external' and 'internal' perspectives, as well as the risk that the deeper underlying phenomenon may be overlooked for the sake of simplicity. If phenomena such as reading, daydreaming or dreaming are considered to be absence, based upon their conceptual nature, it would seem that an understanding of presence as a subjective psychological state is no longer salient as it is now determined based on subjective location of the body, regardless of where the mind is focused.

The experience of reading, daydreaming or dreaming poses several complex questions. What occurs when an individual feels as though they are both 'there' as well as 'here'? Is one partially absent, or partially present? The laws of physics state that if a physical entity is real, if they are absent, then equally they are present somewhere. If someone is observed to be deep in thought, or 'absent', from reality, it does not necessarily mean that the experience of the person being observed is that of being absent from the locale of their mind. If, as the above example illustrates, one considers oneself both 'there' as well as 'here', is one *neither* absent *or* present, or are they both absent and present?

In terms of the methodologies used in past presence research studies, it is generally understood that the focus is on the measurement of the *effects of presence*, rather than the *experience of presence*. Presence in itself is a phenomenon that is felt through the physical body as well as the mind, and focussing on one aspect over another may not provide an authentic' understanding of the overall experience. Having these qualities, this phenomenon requires the understanding of not only the *quantity* of the experience, but also the *quality* of the experience.

Whilst the current focus of presence research is on the cognitive and experiential aspects of presence, there still remains a significant need for discussion linking cognitive activities of the design process with research such as presence, specifically, non-technology mediated presence. The research outlined in this book responds to this need. The review of presence has revealed opportunities for a deeper exploration of imagining and the design process.

Chapter 4: Into the study: a Methodology

Introduction

As indicated previously, this book project was prompted by the desire to better understand how designers design, and to explore the implications of this for education in the spatial design disciplines such as architecture and interior design. Of particular interest was the role of imagining in the synthesis stage of the design process. This chapter provides a brief description of the methodological approach and processes undertaken in exploring the phenomenon of imagining. It commences with an overview of Grounded Theory as the methodological framework underpinning and guiding the study.

Grounded Theory as a Methodological Framework

Unlike many research projects that begin with a specific question or hypothesis, this study began with some very broad questions and assumptions. As already highlighted, the questions concerned the nature of imagining, how it is experienced and understood by designers, and how the knowledge on presence generated outside the spatial design areas might inform a richer theoretical understanding of designing. In this respect, the project commenced with three main assumptions: that knowledge to do with imagining could be extended through empirical research involving designers, that this could be further informed by extant theory on presence, and that the resulting expanded theory on imagining would be relevant for design education and practice. At the heart of this was the problem of not knowing what happens in the designer's head in the early stages of design when ideas and proposals are generated. As a design educator, early questions grappled with included: What was it that the students were experiencing? What was going on in the students' mind during the activity of designing? Why was there an apparent 'disconnect' with the surrounding environment during this process? Furthermore, what can be articulated about this process? What can be learnt about this mental process? How is it possible to teach students about a process that is so difficult to articulate?

Not wishing to be too selective or prescriptive at the outset, Grounded Theory was selected as the underpinning and informing methodology, as "Grounded Theory does not spontaneously arise; rather it is generated, developed and integrated by the researcher through the application of the essential Grounded Theory methods" (Birks & Mills, 2011, p. 20).

Data collection – selecting and recruiting data sources

There were five distinct stages of data collection within this study. The first three stages were vital in aiming to understand the phenomenon of imagining during the abductive-synthesis stage in the design process, a significantly complex activity. As data collection is a highly specialised process, it was also necessary to learn *what* to ask, and *how* to ask the correct questions to elicit the required information on a certain phenomenon. Additionally, the early stages of data collection (stages 1-3) required the researcher to examine and consider a variety of evaluation methods and data sources, planning the study and the methods employed. With each stage, it was important to revisit and carefully evaluate the questions and which methods could actually provide the answers to the given questions. As the researcher was seeking to understand *how designers design* and exploring the right questions to ask participants as well as choosing the correct method of data collection and analysis, stages one to three were considered exploratory studies only; the outcomes were used to inform the data collection methods in two stages of the primary study. The planning of the method of data collection and analysis during stages one to three enabled the researcher to learn techniques to utilise time in a more effective and efficient way, develop a new and critical way of thinking through the use of sifting and sorting techniques, and inform the understanding of the phenomenon through the exploration of diverse sources. Figure 4.4 illustrates the five stages of the study.

For stage one of the exploratory data collection, 42 students from a three-year university interior design course participated. Although the data from this study were not intended to be formally included in this book, an ethical clearance was deemed necessary and was undertaken according to the relevant ethics guidelines.

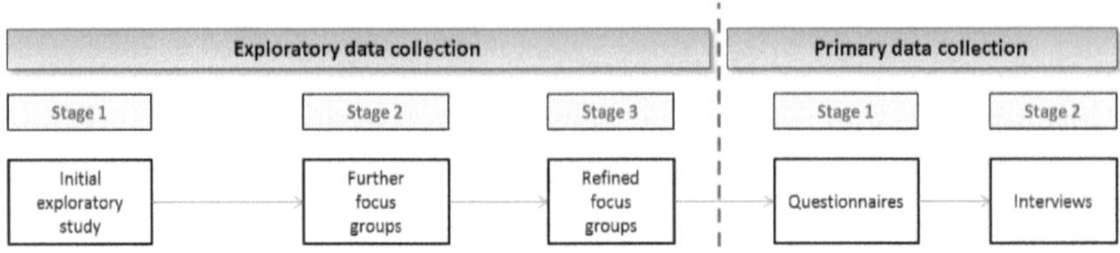

Figure 4.4: The five stages in this study

Participants were recruited through email and each participant was given the option to consider if they wanted to take part in this study; a total of 60 students were invited to participate and 42 students agreed to take part. The 18 students who did not wish to participate in the study did not cite a reason. Of the 42 students who did participate, all were undertaking Interior Design Interior Design 4, a core interior design unit delivered in second semester of second year in a three year degree. There were 34 females and eight males with less than 10% of the class working within design practice or having had experience working within design practice. The results of this exploratory study from these students served as the 'backbone' for this study, but as stated above, the results are not discussed in this book. Data were collected through several unstructured group interviews that centred on imagining and designing. An example of the questions is provided in the exploratory data collection – the semi-structured interview section below.

When the data through this initial purposive sampling were analysed, it became apparent that more information was needed to saturate the emerging categories. Consequently, stage two of the exploratory data collection was undertaken involving four focus groups of five students in each group nearing completion of their degree (Figure 4.5). Participants for stage two were recruited in the same manner as stage one, discussed above. Twenty-eight students were invited to participate with a total of 20 students agreeing to contribute to the study. The eight students who did not wish to participate in the study cited being busy as a reason. Of the 20 students who did participate, all were undertaking their final year studies in a four year design degree majoring in interior design and architecture. There were 14 females and six males, with more than 85% having had design practice experience.

Making strategic decisions such as this about what or who will provide the richest data constitutes what is known in Grounded Theory as theoretical sampling. The focus group data from stage two were then analysed as part of the constant comparative analysis method further highlighting the need for additional new sources of data. For stage three of the exploratory data collection, the source of data was opened up to include five landscape students as well as five architecture students undertaking postgraduate study who were recruited in the same manner as in study one and two. Of the 10 students who did participate, there were two females and eight males, with all having had at least three months design practice experience.

The data collected through these focus groups were analysed using an intermediate level of coding and ultimately informed the development of the primary data collection. For stage one of this process, potential participants were sent a questionnaire by email and of the 482 people who were sent questionnaires, 54 individuals responded and returned completed questionnaires. The data were then subjected to advanced coding and theoretical integration with the second and final stage of the primary data collection; the interviews providing theoretical saturation for the study.

In order to source participants for stage two of the primary data collection, approximately 45 architects and interior designers from South East Queensland were emailed and telephoned regarding the study and were invited to participate. Of these, 24 agreed to the study but ten participants withdrew at the last minute citing work commitments. Following a further two last minute cancellations by participants, fourteen interviews were undertaken and 10 interviews were transcribed; four were not transcribed due to very poor audio quality due to the environment they were recorded within. Further efforts were made to source more participants but potential participants refused citing work commitments or agreed to the interview, then cancelled. Respondents also took an average of three weeks up to more than a month to respond to requests for interviews, thus constraints of time and availability of subjects prevented the gathering of more data for this study. Only those who had a minimum of five years' experience with several completed projects were selected for the study because the interviewer required information about the initial aspect of their personal design process, the design process prior to the completed project and finally, the comparison afterward: was it similar to the space that they had pictured? Or was it completely different to the space that they had imagined? The intention was that if they had gone as far as achieving a 'deeper' imagining experience during the design process, the interviewer was interested to see whether the place they had 'been to' was different from the completed space, and if so, how different was it in terms of look and the feel of the space. It was also important that the projects that were to be discussed were completely new spaces, not those the participants had been to before. It was important that they were not recalling a space that they were familiar with, but rather, a completely new space. It was seen that there was a potential that memory could assist in certain experiences of the space they had been to previously, and this might cloud the experience and affect the outcome or their recollections.

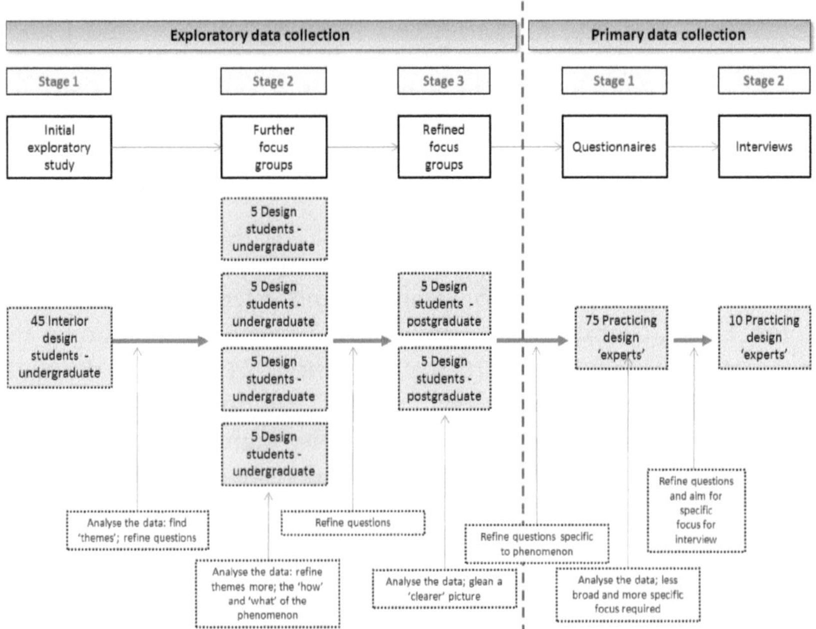

Figure 4.5: Sampling in this study

The theoretical sampling assisted in the development, elaboration and refinement of categories of imagining as described by participants in this study through searching uncovered properties and dimensions of imagining, until no other were found. This process was not designed for searching for repeatedly occurring properties or dimensions of a category (Charmaz, 2006), but rather, was aimed at increasing the range of variation and analytic density of the categories, thus placing them on firmer and stronger theoretical grounds (Strauss & Corbin, 1998). Over the course of these stages, the researcher became more immersed in the data with the level of theoretical sensitivity to analytical possibilities increasing.

Exploratory data collection – the initial semi-structured interview (Stage 1)

A semi-structured interview was chosen for the exploratory study for greater flexibility and potential to yield rich and varied data through the participants playing a central role in shaping the direction of the interview. This is in comparison to a structured interview where the questions are short and rather specific and tend to appear formal to the participants. Questions asked included:

- Do you experience a sense of 'being present' whilst within the 'act of creation' of your interior space?
- How could you describe your sense of 'being present' (experience) that you experience within your virtual interior environment?
- How does this inform the design of your physical interior environment?
- What do you believe sense of 'being present' to be, thinking about the moment when you are immersed in the creation of your space?
- When in the act of designing this 'space', do you 'engage' emotionally with the space in the virtual environment?
- Do you believe that immersion, or the perception of being enveloped, plays a part in experiencing sense of 'being present'?
- Do you believe that sense of 'being present' results from a simple displacement of attention from the real world to the virtual environment, or must one become totally involved in the virtual environment to experience presence?
- Do you perceive yourself to be within your virtual space, or outside of it?
- Are you 'lost' in virtual space in much the same way as you are lost in thought?

- Are you considering the experience of the space whilst you are creating it?
- Do you think it's important that we experience the sequences of experiences one can have within interior spaces? Does it matter or not?
- Do you believe that these experiences are those that satisfy human nature, or is the virtual environment a contrived environment that is far removed from the real world?
- Does interior architecture and design account for sequences of experiences, or the sense of sense of 'being present' within our built environments?
- Do you believe it necessary that these experiences are those that satisfy human nature, and even create a sense of the poetic? If so, why? If not, why not?
- Do you believe that is it important, or even necessary, that we apply the concept of sense of 'being present' to interior architecture, in the context of the built environment?
- Is it significant that our real environments are far removed (in terms of experiential translation) from contrived three-dimensional digital environments?
- Do you believe that there is a 'merging' of self and 'place', in interior architecture and design spaces today?

When the data were analysed from this initial stage the results indicated that the use of a questionnaire was valuable in terms of gathering data on imagining, however, as the students were not experienced designers, the results of the survey did not provide enough in-depth information about the abduction-synthesis stage in designing. As such, the participants were not able to sufficiently articulate their process, nor were they able to adequately reflect on their own design process. Furthermore, there were too many questions posed, thus causing the participants to 'lose track' of the line of questioning.

Exploratory data collection – the focus groups (Stages 2 and 3)

As the researcher was trialling the best methods of data collection in order to understand the phenomenon, rather than a semi-structured interview approach used in stage one, focus groups were used as a method of data collection for stages two and three. According to

Kreuger (1988) focus groups can be used at the preliminary or exploratory stages of a study, can assist in exploring or generating hypotheses (Powell & Single 1996), and are also useful in developing questions or concepts for questionnaires and interview guides (Hoppe, Wells, Morrison, Gilmore & Wilsdon, 1995; Lankshear 1993).

As the primary purpose of focus group research is to draw upon the participant's feelings, attitudes, beliefs reactions and feelings in a way which would not be feasible using other methods (e.g. observation, one-to-one interviewing, or questionnaires), it was seen that focus groups would be the best method for data collection in stages two and three. The questions asked in the focus groups for stage two included:

- Could you describe what happens when start designing something?
- When you are designing, what do you experience?
- When does it occur?
- How often does it occur?
- What increases the experience?
- What is it like?
- Do you perceive yourself to be within your imagined environment, or outside of it? Can you explain this?
- Finally, could you describe the difference of experiences between the first two questions?

The participants were also asked to rate on a scale of 1-7 with 1= not at all; 7 = very much:

- When you are designing, to what extent do you feel like you are inside the environment?
- To what extent do you feel immersed in the environment?
- To what extent do you feel surrounded by your (imagined) environment?
- To what extent do you feel submerged in your (imagined) environment?
- When designing, do you experience a sense of being 'really there' inside the virtual environment? That is, do you

experience a sense of 'being present in the space' whilst within the 'act of creation' of your interior space?

- Do you think that this could potentially inform the design of your physical interior environment?

The second focus group (stage 3) was a smaller group of students who were asked to provide their feedback on a further refined survey. However, this survey differed from stage two as the researcher trialled an edited form of the Presence Questionnaire (PQ) (Witmer & Singer, 1998). However, when analysing the data the results indicated that the use of quantitative methods did not adequately bring out the richness of data that qualitative methods can do when exploring a phenomenon such as imagining. The participants were asked to respond to a series of questions

In terms of the questionnaire data, these were analysed using manual tools, and then re-analysed using Leximancer. MaxQDA, a qualitative software package, which was initially trialled for entering codes. However, the use of MaxQDA was found to be hindering the process of becoming deeply familiar with the data and thus only manual analysis continued. The reasoning for the use of manual analysis over the use of digital tools was so the researcher could become more immersed in the data by noting, commenting and memoing on each of the participants' responses. Analysis of the data began when all questionnaires were returned and after the first interview was transcribed, and the analysis continued throughout the data collection period using: *open coding*, where the data was examined to identify relevant categories; *axial coding*, where the categories were refined, developed and related; and finally, *selective coding*, where the central category was identified and related to other categories. To remain theoretically sensitive the research kept a set of questions at the forefront when undertaking the different levels of coding: 'what is this data a study of?'; 'what category of codes does this event indicate?' and finally, 'what is happening here in the data?' According to Glaser (1978) and Charmaz (2006), these questions greatly assist the researcher in keeping theoretically sensitive and focussed when collecting, coding and analysing data.

Summary

This chapter has presented a brief description of the methodological approach and process undertaken in exploring the phenomenon of imagining. As explained, Grounded Theory methodology was selected to inform and guide the study because of its strong philosophical alignment with the research context and its aims and objectives. In line with 'good practice' in Grounded Theory, the chapter has attempted to as transparent as possible, outlining in detail the research journey and processes of data collection management and analysis. The next two chapters present the outcomes of those processes in the form of a Taxonomy of Imagining and the Spatial Design Imagining (SDI) Model, representing the substantive theory emerging from the 'ground' of participant experiences and research interpretation.

Chapter 5: Into the Study: Empirically Grounded Imagining

Introduction

This chapter focuses on the outcome of analysis of empirical data obtained from the participants of the study when describing their experiences of imagining in the spatial design disciplines. The first section describes how designers regard imagining in the context of the design process. The second section is the outcome of a closer scrutiny of imagining and the development of empirically grounded categories of imagining. The third section provides an outline of what designers consider to be the main enablers of, and barriers to, imagining. When these are considered in the following chapter in relation to the outcome of the exploration of presence and design methodology literature, a theoretical foundation is established for speculating about a new theory of imagining in spatial designing in the form of the Spatial Design Imagining (SDI) Model.

Imagining and Designing

This section highlights connections between participants' experience of designing and their understanding of imagining. For several designer participants, in the initial stages, designing is regarded as a highly qualitative and speculative process that oscillates between the brain and various external spaces, whether they are tangible spaces such as pieces of paper or virtual spaces produced by digital software such as CAD (Computer Aided Design). They describe how ideas develop in both spaces but in different ways. While the externalised ideas are explored in an attentive and focussed way, their inception and further development in the brain is more unconscious or subconscious. Ideas are intentionally left to incubate before they are externalised and developed further in a more direct and conscious way. Integral to this is imagination and how it "allows you to see with the mind's eye what has not yet been drawn" [wl]. As conveyed in the following response, the respondent believes that some people have the ability to do this effectively others do not and that the process takes place in its own time. When it is forced it is understood to lose its inspirational quality.

> *Really hard to quantify...In architecture it involves lots of space planning, head scratching, scribbling, getting ideas down on paper (or CAD), then letting it rattle around in my head for a while, then working at it again. It is something that just happens, you either have it or you don't. It can be forced, and that is often evident in the end result, which is often lacking in inspiration [br].*

The need to allow an idea to gestate is described by several other designers as follows:

> *When I start a design process I just begin with scribbling and sketching out ideas. As soon as I think of one shape - form - concept etc another idea will pop into my head. I will keep drawing, researching and searching until I get an idea that I am fond of. Once I have this idea it is hard for me to stop thinking about it - I have to keep drawing it, from different angles and different shapes until the idea is fleshed out. Once I have the idea, I am content and can relax. I will leave it a couple of days then come back to it and re-think the idea, I will seek feedback... this will normally point me in a new direction [mg].*

> *My way of working is quite intuitive. I read widely and inform myself about all facets of the project, especially the social and cultural values of the community who will use my building. Then I leave this to accumulate unordered in my subconscious. Once I know all the hidden facets of the project as well as the functional requirements, I conceive a coherent spatial whole that brings together the diverse elements. This thinking often occurs in the early hours of the morning, and I have trained myself to think in precise spatial terms from the outset. If I can mentally conceive an answer, I am content to carry it around in my mind, which allows for further distillation of the embryonic solution... Since all this happens at a subconscious level, it is not clear to me precisely how it comes together … it just does [mn].*

The preceding statements also reveal the interplay of different forms of cognition involving imagining, induction and deduction. In the following response, imagining is understood to play a central role in reflection and the development of a mental model providing the basis for externalised (physical or digital) modelling. This follows:

> *...the pragmatic stage of reading and understanding the brief first. Then there is some period of contemplation, arranging the spaces in my mind, developing a model concept. This is usually assisted with the use of freehand sketches, looking at photographs, visiting or revisiting other sites and recalling events, characteristics and emotions associated with other projects, both real and illusionary. Next is the start of committing to paper and then to computer modelling combined with continually adjusting the spaces and trying different sizes shapes or patterns to solidify the illusion. Finally the nuts and bolts part, the fine tuning, squeezing retrying layouts, exploring further layouts and moulding the shape, materials and finishes to achieve and visualise the concept, revisit the design brief check the regulatory requirements and finalise the computer model [pt].*

For another designer:

> *Initial analysis of constraints and brief leads to a period of free form speculation from which one alternative is selected, sometimes arbitrarily as the matrix for exploring other aspects; structural, environmental, servicing and impact. Any of these aspects and others may prompt reconsideration of alternatives or refinement of the one used for initial investigation; at this stage it is important not to become fixed on one solution, which might block consideration of others. At this stage an attempt is made to express the site potential and the results of the brief; the exercise of explaining these aspects to clients usually firms up priorities. Then the 'analysis-synthesis loop' is followed until it is clear that one solution can meet the objectives in a way that optimises the site and resources available [dk].*

Providing additional information, the following response conveys how designing also involves establishing constraints as part of defining a problem space for imagining and/or deducing potential solutions:

> *...Establishing parameters, imagining and/or deducing potential solutions, conceptualising these in built form, testing/review of assumptions/decisions made against outcome desired/required. Always being open to the unexpected and not discarding the 'out of left field' solutions or ideas without analysis [gr].*

In addition, it reveals an appreciation of the need to translate ideas into built form providing a less abstract, more 'real' basis for reviewing assumptions as well as the decisions. The use of the term 'review' in conjunction with 'testing' suggests an appreciation of the qualitative nature of designing, with the outputs from imagining and deduction needing to be challenged and checked against desires as well as requirements. Overall, the process is regarded as speculative with outcomes regarded tentatively in terms of their potential. Even though the outcomes might be unexpected and emerge in unlikely ways, the participant recommends being prepared to at least analyse them first rather than ignoring them.

While the response displays hesitation in too quickly accepting an outcome as the final outcome, it also recognises the need to have a sense of whole early in the process in order to immerse oneself in it and vicariously develop further appreciation of the issues by acting out roles of those who have a connection to it. In this respect, imagination is understood to play a significant heuristic role.

> *At this stage you are trying to refrain from imposing the solution prematurely but you need to be able to place yourself within the overall context of the problem in order to form views of what the issues are and make sense of the project. You need for example at various times to "be" the client, the local authority town planner, the lending authority, the potential buyer, the builder, the tradesman, the neighbour, etc. Your imagination is what allows you to do this [mo].*

Designing as involving a balancing act between the need to take action in an imaginative way and the need to be cautious and critical is conveyed as follows:

> *I think imagining plays a part, to some extent. It is important because it allows the creative process to start flowing. At the same time, it can be dangerous to start imagining a finished*

> *product this early in the process because there is the possibility that you will not be able to see past that original idea. It is important that you keep an open mind while designing and not get too attached to a particular design because it is always possible for a design to be worked further. Although imagining can be dangerous, it also helps to encourage you to design. When you can imagine the finished product or even a part of it, it helps to motivate you and connect you to the design [an].*

And from other designers: "when managed or used well it [imagining] can be enormously productive and is often the main part of the amazing solution" [an], however, there still remains the need to balance between "the logistical and scientific aspects otherwise architecture is just a fashion statement and a neglect of duty by the architect" [ck].

Throughout the entire process of design, from the beginning to the conclusion of the project, designing is very much involved with relationships and processes during different stages, with respondents maintaining that the design process "is founded on the relationship between client, design consultant and the wider world, but it's also about what the designer can bring to the table and how he [sic] brings it to the table (politics of persuasion)" [by]. "The two stages are enmeshed – it is the play between the idea and the possibility of the idea(s) that meets the client's requirements on the one part and the designer's possibility on the other" [tk].

One designer describes the initial design process as 'synthesis':

> *...synthesis of conceptual solutions to the issues [are] identified during the prior phase of understanding of the brief / definition of the design parameters. This synthesis is achieved through an iterative, evolutionary process, bringing to bear the capability of the designer to manage and apply their range of knowledge, experience and skill, to develop potential solutions, and then evaluate them against criteria established during the briefing phase, before moving to the next iteration, keeping the positive aspects, and creating additional ideas to replace the negative aspects, and so –on until the design stands scrutiny against all relevant criteria [rn].*

As conveyed in early statements, the process of designing is understood as iterative and evolutionary, with the knowledge, experience and skill of the designer being critical. Designers are divided in how they approach the initial stage of design. Some strongly advocate that rather than designing around what is first 'intuited', or 'seen in the inner vision', the activities should rather be about "getting the brief correct". For one designer, it is about:

> *Initially, attaining a good, workable brief with the client, then designing a concept that matches the requirements to the characteristics of the site, its location and its surroundings. Then, with the client, reaffirming the brief, augmenting it as necessary and, in the process judging whether the concept is comprehensive and flexible enough to form the basis of refinement of the design [sh].*

Visiting the site can help in finding an aesthetic theme or 'hook' for the project:

> *It depends on how you approach the site in the beginning; what time of the day can influence the design process. If you go in the morning sunlight may be flooding the site giving you warmth that you want to capture. It may be a view, or the lack of privacy. It is more like capturing a feeling, the essence of what the site says to you that you try to capture in the design [wl].*

> *Site and budget are everything. Architecture is about people's use of buildings and their surrounds, not an end in itself. For clients, begin by looking for a 'hook', an inspiration, an angle that they are passionate about, then push this to an extreme [jt].*

Other strategies involve using design expertise gained though previous experiences and projects. This is tapped into intuitively when working on a new project, particularly informing the questions asked and interpreting subsequent responses:

> *A lot of experience is very helpful in undertaking the initial work because there is a lot of intuition involved in all aspects including asking the right questions to establish the brief,*

> *reading beyond what you are told, designing a concept that works extremely well and also has the potential not only for development into good, appropriate architecture but that also has the capacity, if appropriate, to be changed and augmented with the efflux of time [sh].*

For other designers, there is the suggestion that the design solution will emerge inductively and deductively from the information collected: "to understand the brief and the context in all its aspects including economic, social, regulatory and physical in the belief that the design solution will emerge from these constraints" [mo]. Indeed, as expressed by one designer:

> *I was educated to consider that preconceived ideas of any kind inhibit the chance to achieve very good and even great architecture and I have had every reason to applaud that teaching. Until one has a full brief and full understanding of its implications and a full evaluation of the site and it surroundings and a full understanding of how the regulations governing the development of the site and the development of the category of building being sought will affect the design, any preconceptions are absolutely worse than useless [sd].*

When asked about imagining in their designing, several designers used it in relation to 'visualisation', not only for themselves but also in arriving at a shared understanding with others:

> *I have used creative visualisation successfully across cultural differences with an Indigenous person working with me. It is a very powerful tool to assist people to understand the complexity of the design process and the range of factors and issues we take into consideration in a design process [bk].*

> *The ability to visualise a 3D concept is a huge advantage to have confidence in exploring and examining any one (or a range of) solutions. It also requires the communication skills to convey the visualisation in ways that "connect" with the client and ensure confirmation of effective communication and understanding [jn].*

While many designers use imagining to more effectively engage themselves and others with the design process and what is being designed, there was one designer who used imagining to distance themselves from the building and allow it to design itself:

> *A diversity of sights appear to the "inward eye". This is very distracting. I try, unsuccessfully I am sure, to take my "self" out of the building. I do this by playing a type of game. The game goes something like this: -*
>
> ** The building is designing itself.*
>
> ** Out of all the myriad forces shaping it, including financial, regulatory, aesthetic, contextual, technological, etc. what does the building want to become? Not "what do we want it to become".*
>
> ** I am asking the building what it wants to be?*
>
> ** How will it form itself into an aesthetic entity whilst fulfilling all of these competing requirements? [mo]*

In summary, the empirical data reveals that the designers participating in the study understand designing as heuristic, iterative and evolutionary. They describe the process as demanding particular ways of working in a space between the external physical world and the internal imagined world. Negotiating and navigating these worlds and the space in between involves different ways of thinking.

While induction is discussed in relation to drawing meaning out of information collected, imagining is regarded as crucial in creative mental synthesis. That is, in intuitively developing tentative 'wholes' of varying levels of abstraction as places for conjecture, simulation and modelling, varying degrees and types of immersion and engagement by the designer and others are involved. Sometimes the engagement is conceptual; sometimes it is also emotive and aesthetic in the existential sense. Mental models and mental modelling are linked with facilitating incubation, gestation and distillation at a subconscious level; a process that is understood as highly original and creative. The following section describes the outcome of closer examination of imagining and the various ways this is employed by designers especially in relation to mental modelling and creative mental synthesis.

Empirically Grounded Categories of Imagining

As conveyed in Figure 5.1, the analysis of empirical data produced four categories of imagining: visual imagining (comprising spatial and pictorial imagining); aesthetic imagining; and (con)textual imagining.

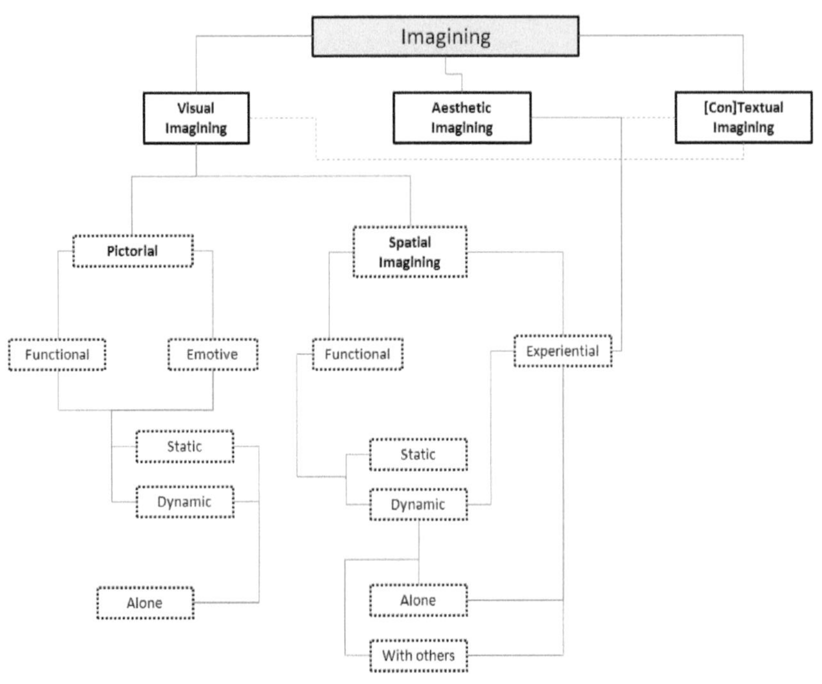

Figure 5.1: Empirically grounded categories of imagining

Visual Imagining

Visual imagining "...allows you to see with the mind's eye what has not yet been drawn" [an].

Pictorial – functional

Several designers discussed their experience of designing giving emphasis to possible ways in which the environment might operate, including how spaces might connect and interconnect to support certain activities. For them, imagining is a 'cognitive plane' where: "internal spaces are visualised and their interrelationship with other interior spaces is conceived in my mind" [mn] as a two dimensional (2D) or pictorial image such as a floor plan, wall elevation or section:

> *The ability to visualise a built form from a plan view develops over time. This especially applies when the plan and or section are complex. The ability to read architectural drawings is probably much more restricted than is generally recognised. I work on plan view mainly, but with the section in mind. When the section is the generator I work more in section [ja].*

In this category, the translation into 3D is digital rather than cognitive:

> *One of the positives of CAD is the ability to generate a model at an early stage. Most of my design development and documentation is now done in CAD, working in 3D, where the model is the source of all views, including walk-throughs, fly-overs, etc [ja].*

Similarly for another designer, various concrete elements are the starting point: "roof and wall envelope shape is the starting point, particulars, eg windows, doors, balustrades get added later" [kz]. Where images do have a 3D quality, these are mechanical, described as: "an image usually three dimensional i.e. photorealistic, sometimes more geometric, rather like an isometric drawing, sometimes a view from a long way away so as to imagine the overall form of the structure in two dimensions, like an elevation drawing" [bn].

In terms of the relationship between the designer and the 'object' of view, this can be static as when working mainly with plans and elevations; or dynamic where the 2D views represent different orientations between the designer and the object, or where the objects have been rotated in relation to the designer.

<u>Pictorial - emotive</u>

Sometimes these descriptions included reference to formal qualities and the possible finishes: "Visualisation (in the form of creating mental images) occurs in reviewing possible room layouts, external and internal form and material selections as part of the production of concept layout/s" [bn] suggesting the intent to convey a particular feel or mood.

Spatial – functional

As indicated in the previous section, several designers discussed their experience of designing giving emphasis to possible ways in which the environment might operate, including how spaces might connect and interconnect to support certain activities. For these designers, their relationship with the object was essentially two dimensional or pictorial. For others however, what is visualised has a three dimensional (3D) or spatial quality: Spatial elements are explored "to investigate the possibilities of the space, and develop relationships between the spaces. It can often be easier to visualise something in 3D than to try and draw it. It gets lost in the translation" [mk].

As indicated in the previous quotation, 3D visualisation is discussed in relation to the externalisation process of drawing and how this latter process on its own is less successful than first attempting to form a 3D image in the mind. In this respect, and as also conveyed in the following quotations, 3D imagining is a powerful translational strategy: "Without this [3D] imagination there would be no foundation for further design. Researching facts could be one way but the research always needs an imagination for it to be translated to a physical design" [wk]. "it is the driving force behind realisation of the brief. If you have no imagination then you cannot design and you cannot provide a solution for the client" [ck]. Added to this is the role of imagining in communication with the client:

> *I designed a house that is totally made of timber, totally contrary to what he said, just because the spot, it had to be attached to another Swiss chalet and knowing him, all he wanted or I thought he wanted was space and volume. So…..and I had the authority of a visible structure two talk at once columns and beams and all that have to be much bigger because of the snow than what they are here. And I saw that vision from the start, large timber columns, timber panel finishing with shadows drawn against these columns. Columns visible, structure is all visible and huge big windows which were the opposite to what he had before. And I had that in my mind when I talked to him about it before doing anything. He said, oh no I totally don't want timber. Okay. I'll show you and I did that, that's when I went straight to computer drew the*

> *picture in 3D…and where the plumbing would be and he was convinced and the house was built like that.*
>
> *I: And the picture that you imagined in your mind's eye is the same as what came out on the computer?*
>
> *[sg]; It's exactly that.*

While some designers appear to employ spatial imagining in a detached and static way, others are more involved and iteratively engaged. Here spatial imagining is not only used to see the space in the mind's eye as an observer, but also allows the designer to "be in the space", to experience it and test how things may work by "trying things out in your head, thinking about them" [br].

> *[so] I understand imagination to be an aspect of mind. (Not separate.) The thinking mind can know and learn many things but is easily overwhelmed with too much data. By taking time to "be" with the place, the brief and ponder opportunities and challenges inherent in the design situation, one gives one's entire mind – conscious and sub-conscious – to mull over the plethora of issues present in the situation. One may then "intuit" a solution or a range of possible solutions, and "imagine" them – picture them in one's mind or transfer images to sketch paper and "test" them – try them on to see if they fit... This is all accomplished via imagery, which, before it is committed to paper of computer graphics, is first "imagined" in the mind of the architect [so].*

In this category, the purpose of spatial imagining is very much functional and operational:

> *The role of visualisation is to be able to see in your mind how the spaces and the requirements that go on in them, will work and to consider the effects of constraints and other aspects of the project, to minimise conflicts. It is also necessary to visualise the way in which a building will be physically constructed when creating spaces. No point in creating something that won't be achievable [bl]… I picture myself being the end user and using the spaces I am create [in order to] try to see better ways for processes and space relationships, etc., to increase efficiencies and eliminate crossovers [bl].*

While some designers conveyed somewhat simplistic notions of spatial imagining, others were more sophisticated, describing what is imagined as a: "complex overlayering of systems; circulation, habitation, structure, services and sometimes the picture is completed by colours of furnishings [dk].

Where there were references to how the spaces feel, these tended to relate to physical aspects of comfort: "I am within the rooms I am creating and how it may feel to be using those spaces" [jv]

> *The building form, its layout, its circulation patterns, its spaces, its outlook, its potential for being well suited to good placement of furniture and objects and, finally, its appearance. This is not necessarily a complete list. Each factor tends strongly to affect the design of the others...The relationships of the floor levels of the proposed structure to one another and to the levels of the ground, along with other similar considerations, are often major factors in one's mind when establishing principles of design that will be applicable to the specific project being considered [sd].*

According to one of the designers, cognitive 3D modelling is refined by building actual models:

> *Generally the first visual manifestation of a design occurs over a weekend, the only time when the phone stops and I have uninterrupted periods in which to think. I begin by drawing a large-scale isometric or axonometric projection, which evolves naturally as I resolve the most important spaces first. Detailed functional relationships sort themselves out through the process of drawing. Or else, more recently I am building balsa and cardboard models as the first full-blown expression of the idea. There is no freehand sketching in the accepted sense. 'Big ideas' evolve as fully-resolved entities and they rarely change in any dramatic way, unless the functions which underpin the project are revised [mn].*

For another designer:

> *We place into the mixing pot the clients brief, our experiences of the site and its surrounds, thermal considerations and initial thoughts on massing. The butter paper comes out and a lot of sketching and model making occurs until the form starts to take*

> *shape. Once the form and plan has taken shape, we move to 3D CAD. The design is then developed in 3D CAD" [dw].*

While most descriptions placed the designer alone in the imagined spaces, in some instances designers also imagined the clients and others in the spaces undertaking particular activities:

> *Imagining, undertaking activities, often happens with spaces such as kitchens, particularly imagining moving and undertaking activities within or to/from the space, ie. Imagining the suitability of the relationship between several spaces such as kitchen and pantry, or getting something from the fridge, or kitchen and adjacent living space, in which case imagining communicating with guests often occurs [bn].*

<u>Spatial – experiential</u>

As described and illustrated in the previous section, some designers employed spatial imagining in a very functional way, conceptually and substantively. There were other instances however, where the engagement with the imagined spaces extended beyond the functional and where images were understood to facilitate deeper engagement for and with the client; for explaining as well as understanding: "Absolutely important always draw the perspectives- as then I can "see" and the client can see, very few clients can read plans the one that do are engineers or work with plans" [hk]. "A picture says a thousand words. We encourage clients to provide images of works that they like and don't like. This helps to quickly identify a direction of design" [wl]. "From the massing of the design through to the material textures that may be engaged. Whether it is sketched elevations, perspectives, models or 3D CAD, the visual is an important part of understanding and explaining the building" [dw]. Without imagining and "the ability to 3-dimensionally picture an environment and put yourself in the "experience" of it, you cannot fully appreciate what the design solution is" [dn].

One of the consequences of failing to acknowledge an experiential role for imagining may be that the space will end up lacking certain experiential qualities. For "Without imagination it would not be possible to produce a solution that meets the clients brief while producing spaces and buildings that are aesthetically pleasing while enhancing the users experience" [bb].

Unlike in the previous category where the designers were at times detached (even outside the space, such as when viewing the layout of a building from above), in this category the designers invariably are immersed in the space or imagining others there. In this regard, the space becomes a place:

> *[fx]: it's the way I have to do it, is actually engaging with those people and it's because of the way I am, it's just this kind of trying to understand other people and trying to establish a relationship with them and really, really trying to understand who they are. So I'm always placing them, placing them in that. I'm imagining their activities in it. I'm imagining they're in it, not necessarily just me. I'm kind of trying to understand the essence of them and shape these places around them as well so it's all that whole process of kind of pulling all of those things together".*

If the designers are designing a building their immersion in the environment can be outside as well as inside the building: "I am always in the space! The every level is always as could be really seen from a road, or a yard- nothing else is important" [hk]. In all, the designer's relationship with the building is very dynamic: "I also like to see the building as you might from, oddly enough, a helicopter. I often do many sketched aerial views which show the building almost as a sculpture in the street but seen from above" [mo]. For another designer, they describe: "Walking around it, flying over it, and through it; trying to be there and experience it firsthand" [pt]. For another:

> *I am everywhere I need to be. If attempting to appreciate the way a stair meets an upper level landing I may be in mid air beside the subject point say 3 or 400 mm off the stair slightly above or beside. If investigating the meeting of several floor treatments, it may be from directly above and may have an element of zoom. If a complete building it may be from all around including aerially but unlikely underground. Few digital tools have achieved this capability. SketchUp goes some way toward it. ViewBuild is closer in that if applying for example a decorative element you can zoom down through walls and up to the ceiling corner it may be placed. That is very much the way we visualise.*

> *Visualisation (mentally) is vastly more powerful than current tools achieve. Instantaneously you can invest the image with the mien of last night's theatre, a childhood dreaded place or the lofty daydream of a walk among the high eucalypts of early morning [ja].*

The dynamic quality of this relationship is particularly evident in the following response:

> *I imagine myself observing and moving through at many different points of the entire building being created. The Architect must be able to walk through the entire building from the human perspective eye height, but simultaneously be able to view from bird's eye in the sky and at all "visual approaches" to and from the building. This allows full understanding of the dynamic interactive sculpture being created. I also imagine turning the building around in my mind to view different angles at will from my one view point (a bit like rotating a model in your hands) [gg].*

The imagined experience for these designers is also sensorially rich:

> *…I always try to imagine how will it feels being in the space that I design, how I will move, how I will interact with other users, how big is the space, how the light will enter… [og].*

> *…touching, tasting and smelling etc. Using the senses as much as possible [bk]*

> *[cw]: Yeah you can see colours, you can… it's quite eerie because it's a bit spooky I guess but you do walk through it and um… colours and lights on a spatial sense is usually a little bit warped, but yeah you do.*

> *I: Do you feel temperature or light and dark?*

> *[cw]: Light and dark, yeah. Breeze, even things like… not breeze but openness, a sense of openness, there's a sense of that, yeah. You get that feeling walk into a room that has an open side you get a sense of that, you get that… you know? Or the, or if a place is um, more condensed, it's condensed down and you get a sense of that enclosure.*

To further contextualise and enrich the experience, designers visit and experience the site:

> *The outcome develops as you find out what the requirements are and it changes as the requirements are refined. With a house site, I generally walk onto a site and imagine the location of spaces that are usually predetermined by orientation, breezes, views, access, neighbours, inlook, outlook, etc., - all fairly instinctive [bl].*

For some designers, this experiential form of imaging can be so powerful as to inform the design and construction of environments that, when experienced by the designers, feel very familiar:

> *[hc]: That's how I always say....it's umm....I am constantly surprised at how similar they are…I don't get a feeling of newness when I walk in it…truthfully, if you do it well, it shouldn't be a surprise, if you have walked through it already.*

For others, the outcome may have similarities but has evolved further due to other influences: "What I imagine at the beginning of a project and how it turns out in the end are usually very different. Some aspects may remain the same but usually the design has progressed so much from where I started that it is nowhere near what I imagined (usually it's much better)" [gd]. Other respondents feel as though the end outcome is "Better, fully fleshed in all views" [md].

Aesthetic Imagining

> *"Without imagination it would not be possible to produce a solution that meets the clients brief while producing spaces and buildings that are aesthetically pleasing while enhancing the users experience" [an].*

While this category has qualities in common with the previous categories, it also has several elements that differentiate it from the other categories. For experiences exhibiting aesthetic imagining there is recognition of a strong affective relationship with the design project from the start: "you need to be in this space in your mind, you need to clear your mind of all other matters and devote your emotion to the project to get there, otherwise you are just making buildings [jt]. For another designer this involves creating: "an emotional theme for a project in my mind and recall, through my experiences or research new experiences, all the conceptual spaces, forms colours and finishes that equate to that emotion. This then becomes the visualization to target. I don't get it right every time first time" [pt]. For another, the process is also described emotively:

> *The first part of the process is to "feel into" what the project is about. Without the "feel into" process the design is about "due process" which is quite different idiom. From the "feel into" process, the imaginative juices start working through a range of internalised journeys that "make a fit" between the form, the physical requirements of the brief, and a manifestation of the two [tk].*

The emotive dimension also extends to the client: "[I] talk with clients – what accommodation do they want (distinct from what accommodation they need) – how do they live their lives, who cooks, entertain, family life, music, travel, food, drink, art preferences. I want to create a special glove that will embrace their lifestyle" [ig].

In terms of the relationship between the designer and the 'object' of the evolving design, this is very fluid, sometimes to the point where: "I am the object. There is a oneness between the Observer and the Object" [tk]. Or as noted by another designer: "By closing one's eyes and living the building...we try to experience the spaces and use this to communicate the possible experiences to the client" [dw].

The process is also understood in metaphoric and abstract terms: "Far too abstract to make this a place….. I think it is really a swirl of emotion and feeling, like colours half mixed together in a pot, and of course the challenge of architecture is to get this and make it into a built form" [jt]. For some designers, the initial inspiration comes from the site:

> *Sometimes you can just see the building on the site. It depends on how you approach the site in the beginning; what time of the day can influence the design process. If you go in the morning sunlight may be flooding the site giving you warmth that you want to capture. It may be a view, or the lack of privacy. It is more like capturing a feeling, the essence of what the site says to you that you try to capture in the design [wl].*

Also of note in this category is the experiential engagement of the designer, and in some cases, engaging the client experientially:

> *I tend to work kinaesthetically and to feel spaces, especially as they are animated by light, season and time of day. With clients it is often necessary to engage an understanding of size and to set out a space at life size as for instance by pegging out kitchen benchtops and fittings. When the project is an extension or conversion, lines of string may be used to model the proposals at 1:1" [dk].*

In addition, designers use imagining as part of a rigorous process of imaging, representing and testing:

> *"I test it and tune it. This allows you to assess its capacity to engage. Light and sound is probably much more likely than actual temperature except temperature as it may be perceived due to the character of the space" [ja].*

Another differentiating aspect is how designers regard each project as unique: "Every project is different...[in] my mind's eye as you say [I am] focussed on the qualities that I am looking for in this project, be it bold, sympathetic, adventurous, or whatever" [pt]; and the experiential relationship is intimate and all consuming:

> *...the little house we did in Canberra, which was the very first project that we ever got on site, took a bit longer than that to finish, was one of those projects where it wasn't at all about form, it was primarily about space and materials and and...and...light. And it was one of those projects where, and because it was for the family, and I was very sort of intimately connected with it, I really…it was for my father and his partner and I wanted it to be magnificent and that was…and even though they didn't necessarily have this objective of magnificence, they really sort of wanted to take a north eastern corner and carve out a laundry and a toilet and make it useable. I really, you know, you feel like you owe your parents a lot and you do. I really wanted to make this thing that was just a delight for them because they're delightful people, they love to read and drink wine in the afternoon and have great lunches and so on...*

> *Yeah and I knew that they would really enjoy this if I could just make it right, so I spent a lot of time considering what the materials were and how it would orient it towards Mt Andrew which is over there and how the sun would come across at different times of the year. And I could do a lot of banal kind of testing of that in computers and relations but a lot of it was really about just, really just placing yourself there and imagining. Recycled timber, what it feels like and how the glass feels when it's really cold outside and it's warm inside and things like that. And having, and gone there and seen it finished it's just….it's, you walk in to the space and there's just no surprises. You kind of, you've already been there and it's just taken time to just sort of come into existence.*

(Con)textual Imagining

Whereas the designer (and in some instances the client and users) occupy a poetic place in aesthetic imagining, for (con)textual imagining the place or space is a literary one occupied to better understand and develop a pragmatic context for guiding later more abstract and conceptual stages: "the pragmatic stage of reading and understanding the brief [is] first. Then there is some period of contemplation, arranging the spaces in my mind, developing a model concept" [pt]. Similarly for another designer they: "gather available information, identify constraints and opportunities; develop a conceptual framework for design; thumbnail sketches progress to schematic design, with the form constantly reviewed and reinterpreted in order to strengthening the communication of the concept" [ph]. For one designer, a direct reference is made to the need to collect data: "I am aware of the need to collect and analyse data. This deliberately avoids any pre-conceived ideas and is communicated to clients. i.e. let me have a look at what you say you want in your brief, and let me have a look at the site or the existing building etc., and then we will see what comes out of this" [bn].

This category of imagining then is regarded as particularly critical in the initial brief development and design development stages. According to one designer, the process starts with asking: "What is the question that the design process has to solve? There are too many correct answers to the wrong question. We spend too little time on defining the issues and instead start designing around preconceived concepts" [wb]. "[T]he process of design at these stages should incorporate developing an understanding of the client's needs and wants, as well as an understanding of the possibilities of the site and the immediate environment" [mk].

In later stages, the designer may adopt various roles to challenge proposals:

> *...you are trying to refrain from imposing the solution prematurely but you need to be able to place yourself within the overall context of the problem in order to form views of what the issues are and make sense of the project. You need for example at various times to "be" the client, the local authority town planner, the lending authority, the potential buyer, the builder, the tradesman, the neighbour, etc. Your imagination is what allows you to do this [mo].*

Overall, understanding the context and applying contextual elements in a process of textual imagining is regarded as critical to the design process:

> *As previously mentioned there is a need to appreciate and assimilate the brief and the myriad constraints relevant to the design problem and to understand the immediate and wider context including economic, social, regulatory and physical factors. As the preliminary designs emerge from these forces they are evaluated and redesigned. This redesign almost invariably involves a re-evaluation of the brief, the design constraints and the context. This circular process is repeated many, many times until the desired design emerges. In the case of a building the process of design, evaluation, re-evaluation and redesign goes on until the last worker leaves the site [mo].*

In summary, this section outlines four categories of imagining emerging from analysis of the empirical data comprising designers' experiences of designing and imagining. These are labelled as: visual imagining (comprising spatial and pictorial imagining); aesthetic imagining; and (con)textual imagining. As the following section illustrates, undertaking imagining in the context of mental modelling demands a particular relationship between the designer and the physical environment in which they are located at that time.

Factors Influencing Imagining

For many designers, designing demands mental and sometimes physical withdrawal from their physical world:

> *You need to zone out and concentrate on the imagination of your mind to fully experience the developing concept. There are too many interruptions in the work environment to not require this to happen [pt]*

Many of the designers interviewed regarded various elements in the environment as being a distraction, and subsequently a negative influence to imagining:

> *I do not like working in open plan offices where there are many distractions to concentration. I like to work in a cloistered environment like a monk [to be able to focus]. This approach is then informed and enriched by having group [team] discussion for review and design [wb].*

> *...very conscious and cannot begin initially at least (ie the very beginning) with any possibility of interruptions [lo].*

As indicated here and in the following statement, several talked about removing themselves from the source of the distraction: "If I have a particularly challenging concept to resolve, I withdraw to a quiet room. I am still aware of my surroundings though" [bl]. Not only is the physical space important, but "Getting away gives the time for inspiration" [br]

For some designers, they are so focussed they become unaware of their surroundings, even to the point of describing their detachment from reality as an 'out of body experience':

> *When deep in the initial concept design phase, this can be a bit of an "out of body experience" as this is thrashing deeply through environments etc in a very fluid manner. I am more aware of the physical environment when in later, more practically fashioned design stages [gg].*

As also conveyed here, this is not so much the case when dealing with more pragmatic aspects of design:

> *...(so) focused on the process that I don't hear what is happening around me. This can happen when I'm drawing or imagining a design. However if I'm trying to read, say regulations I can struggle with the task if the surrounding environment is noisy [wl].*

Some describe a parallel state of being:

> *Others have often noted that I seem totally absorbed at these times. I recognise a heightened focus but would argue that I seldom miss whatever else may be going on around in my immediate environment and do attempt to keep up-to-date with wider situation [ja].*

> *The place in which I am located becomes the background of existence. The sounds and smells are there but I am focusing in my own mind on the visualisation of the task at hand [dw].*

Designers vary in their ability to detach themselves:

> *...always conscious of [the surrounding environment], the interaction of people and spaces and their usage [fw].*

> *At times my environment becomes very intrusive, causing me loss of focus and difficulty with imagination, ideas and visualisation [rn].*

> *I am able to "transport" myself to the design situation and that environment. However removal of distractions and interruptions allow greater freedom and flow of ideas. I prefer a natural environment for greater creativity and have refreshed thinking in or after such renewal [jn].*

For those who can create an imagined world, their time there is not necessarily spent doing concentrated work as noted in the following response:

> *...you become more involved in your sketching, I mean it's more a sort of acknowledgement of them, becomes a subconscious type acknowledgement because you're devoting a lot of attention to your sketching or I mean even CAD, it's really just doodling, so it's not that massive concentration type effort and you're really just doodling and thinking well, put this and put that, doodling with no specific something in mind [eo]*

The need to move awareness back to the physical world is precipitated by the need to externalise the imagined idea:

> *The in process becomes all-engaging - the external environment becomes non-existent, not important, secondary perhaps even tertiary to the inner experience. It is only when the conceptualisation crystallises and the need to externalise the process does the surrounding environment re-engage [tk]*

However, as revealed in the following response, it is possible for a mental state and imagining to be mediated by some aspect of the physical world. Indeed, it appears that the physical process of sketching and what is produced on paper partners with a mental space to form dual spaces for mental engagement:

> *I will lose awareness of time and my physical surroundings. Music will cease without me being aware and I am aware often only of my mental state and the process between my mind and the paper" [ig].*

It appears that when in this mental world, another aspect of the designer's being takes over: "in the 'zone'" – so to speak...my subconscious needs time to put things into place" [so].

As indicated here, and as follows, the space or spaces created have a different temporal quality:

> *At the point of designing I am completely transcended into a different time and space and oblivious of what is happening around me" [bv].*

Designers also describe the need to prepare their mind for imaginative work by engaging in some other activity even another project:

> *If it is difficult to focus, it is usually best to do something else: play some music, walk, eat, rest, read, work on another project for a while....when your mind is ready you know and can get back to productive work [so].*

Conclusion

The findings of this chapter reveal the complex process of imagining and its characteristics. Furthermore, it also reveals that imagining does not begin in a state of 'tabula rasa', or a vacuum, but rather, a designer's evolving understanding of a design situation begins with preconceived aesthetic values, precedents, previous projects and assumptions. It is an interpretative and iterative process that is underlined by an anticipatory projection of meaning.

This chapter presented the outcome of analysis of empirical data obtained from the participants of the study when describing their experiences of designing and imagining in relation to their practice in architecture or interior design. The first section sets the scene by describing how the designers regard designing and, related to this, the phenomenon of imagining. The second section represents the outcome of a closer scrutiny of imagining and the development of empirically grounded categories of imagining. The third section provides an outline of what designers consider to be the main enablers of, and barriers to, imagining. When considered in the following chapter in relation to extant design methodology and presence theory, a theoretical foundation is established for speculating about a new theory of imagining in spatial designing in the form of the Spatial Design Imagining (SDI) Model.

Chapter 6: Into the Study: The Spatial Design Imagining (SDI) model

Introduction

This chapter focusses on the main outcome of the study – the Spatial Design Imagining (SDI) Model. It outlines the development of the model according to the integration of extant theory from design methodology (Chapter 2), the empirically grounded categories of imagining emerging from the research involving design practitioners (Chapter 5), and extant theory from presence research (Chapter 3). The following diagram (Figure 6.1) illustrates the general relationship of the three areas and the connecting role of imagining. It also emphasises the spatial design context as the primary domain and focus for this study. As conveyed in Figure 6.1, the research on presence enables closer examination of imagining, and complements and extends empirical research specifically concerned with imagining as experienced by design practitioners. When considered in relation to existing research on design methodology, the outcome is a enriched understanding of the role of imagining in design and a fertile base for more effectively informing design education, particularly in architecture and interior design.

Imagining and design methodology research

The overview of literature on design methodology presented in Chapter 2 highlights a connection between imagining, design thinking and cognition, and design process (Figure 6.2). The design context is now generally understood as complex, ill-defined and uncertain, and in some respects, 'wicked'. Designers work 'in the now' for projects to be realised in the future. While some projects may be very similar, all projects are unique and novel. Given these qualities of the design context, design process by necessity is heuristic, iterative and satisficing. As noted in Chapter 2, Cross (2001a) draws a direct link between working in novel situations with incomplete information and the need to use imagination and constructive fore-

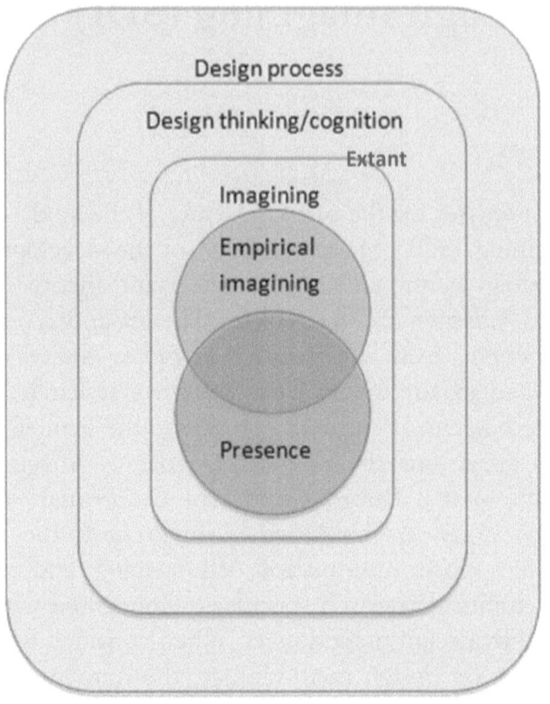

Figure 6.1: Conceptual framework for enriching knowledge of imagining in design

thought in a heuristic way. This realisation led to greater focus in research on design thinking and cognition and thus to an understanding of it being productive and generative, involving conjecture and propositional thinking. This led to further understanding of the central role played by abductive thinking, as opposed to deductive and inductive thinking. According to March (1984), abduction, sometimes also referred to as production and relying in part on intuition, is the only operation which facilitates the development of new ideas; in other words, the process of creation.

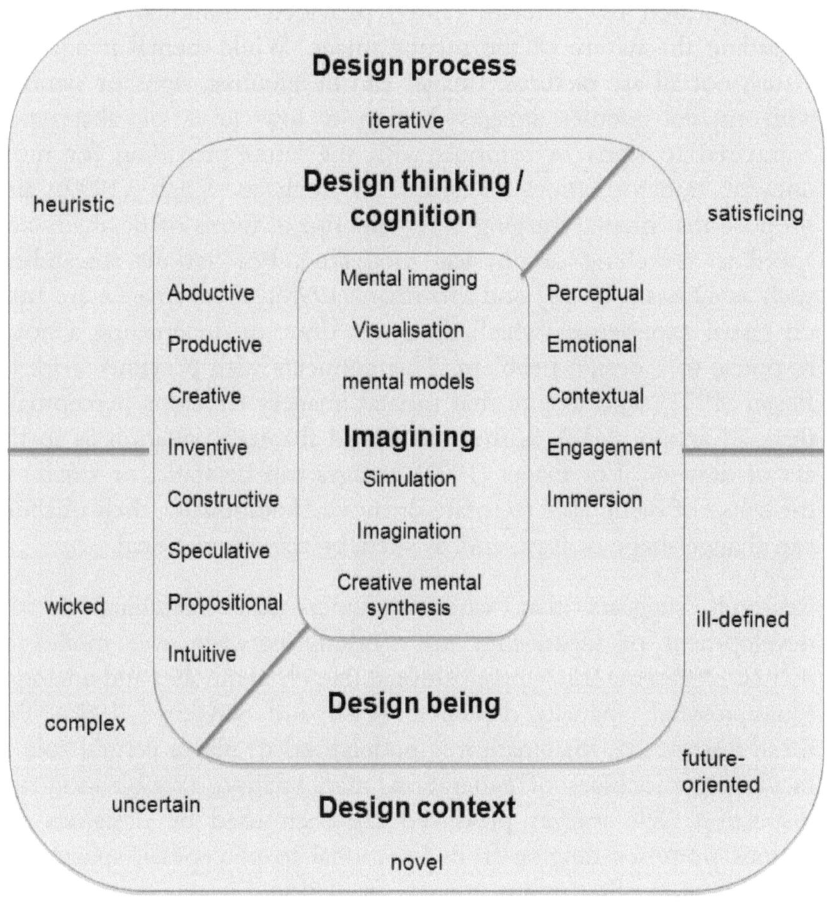

Figure 6.2: Contextual mapping of imagining in design methodology literature.

What this actually involves invites closer examination of the several interconnected concepts highlighted (and sometimes used interchangeably) in design methodology research; concepts such as: imaging, visualisation, mental imaging, mental models, simulation, imagination, and creative mental synthesis. In this study, imaging is accepted to mean the forming of a mental picture of part of the world. By association, these concepts are invariably understood to be visual (as opposed to descriptive) and generated by a process of visualisation or visual thinking. To differentiate them from externalised images, they are also referred to as mental images and the process as mental imaging.

Early research by Arnheim (1969) provided additional infirmation regarding the nature of the mental image. While mental images are visual, not all are pictures. Images can be pictures, signs or symbols with mental picture images having a low level of abstraction compared to signs or symbols with the latter providing for more unusual transformations. Researchers such as Singh (1999) also propose that mental imaging is far superior in terms of flexibility and speed to sketching or physical modelling. For various researchers such as Damaio (1999) and Hoffman (1998), visual images are built on visual experiences which designers draw on in creating a novel response to a design problem. This connects with previous work by Piaget (1971) who asserts that mental imagery develops perceptually through action and is further developed through imitation as in the act of drawing. For Piaget (1971), images can be static, or kinetic in the sense of being able to rotate or move. Additionally, their qualities can change shape or form and, as such, be transformational.

Research suggests that when designers use sketching in the development of ideas, they are working between two modes of mental representation – propositional (largely symbolic) and analogue (quasipictorial, spatially depictive) (Fish and Scrivener, 1990). For these researchers, imagination is understood to play a central role in facilitating processes of generation, manipulation, combination and inspection. The images produced are then used by designers for various purposes ranging from functional to perceptual, speculation and testing; in other words, mental simulation. On this note, research and autobiographies such as that by Frank Lloyd Wright suggest caution in commiting mental images too soon to paper, advocating instead for the conception of the building to live in the imagination and develop some robustness before it is externalised and subjected to external manipulation through such action as sketching.

Mental simulation is considered necessary in designing in order to reduce uncertainty and increase predictive and explanatory power. Mental models help predict through inference how something may function or behave even when the designer has had no previous experience of the situation. While mental models and mental simulation are entirely subjective and imprecise, they have been found to be an effective strategy when working in highly qualitative, indeterminant situations. For Lawson (1980), imagining is integral to mental simulation enabling the designer to draw from their own experience and combine material without any definable (at that time) endpoint in mind.

In this respect, then, designing involves co-evolution of the problem space and the solution space (Maher et al., 1996). Identification of a key concept or 'frame' proceeding to abductive sensemaking are central to this creative synthesising process (Dorst, 2004; Paton & Dorst, 2011). Informing the process is the use of metaphor and analogy, contextual engagement or immersion, and conjecture (Paton & Dorst, 2011).

6.3 Imagining and empirically grounded categories

This section reviews the chapter on empirically grounded categories (Chapter 5) and explores its relationship to imagining as conceptualised in the design methodolology literature reviewed in Chaper 2 and summarised in the previous section. As the discussion reveals, the findings of the empirical study provide additional knowledge about the nature of mental imaging and its relationship to imagining.

Figure 6.3 represents the imagining taxonomy as first presented in Chapter 5. As illustrated, it incorporates three primary categories of imagining - visual imagining, aesthetic imagining and (con)textual imagining, and the two secondary categories of imagining – spatial imagining and pictorial imagining.

In the context of the empirical study informing this taxonomy, visual mental imagining can be pictorial (essentially two dimensional), or spatial (exhibiting more three dimensional realistic qualities). From the designers' experiences it appears that pictorial imagining (such as floor plans or sections) restricts simulation to functional and perceptual aspects, while spatial imagining enables imagined active bodily immersion at a deeper experiential level in the environment being created. In addition to their own experiences, spatial imagining also invites the designer to imagine how others might experience the environment; how it might support psychosocial as well as perceptual, physiological and practical needs.

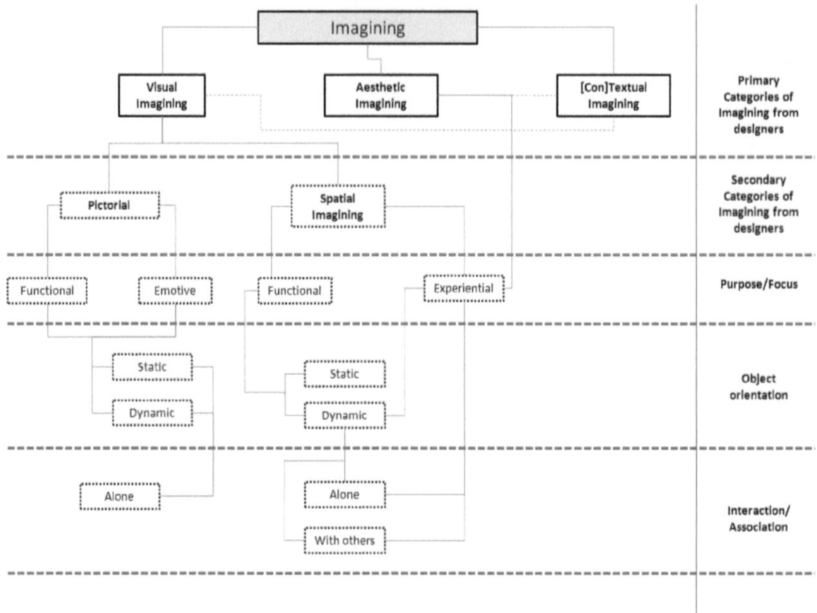

Figure 6.3: Taxonomy of imagining from empirical data

In many cases, the designer augments mental models with externalised models in the form of two-dimensional and three-dimensional drawing and sketching. Increasingly, designers use digitally mediated three-dimensional drawings to aid representation, imagining and testing, with many integrating external physical models (drawings or actual three-dimensional models) with mental models. As noted in the design methodology literature, these models have varying degrees of abstraction ranging from propositional (largely symbolic) and analogue (quasipictorial, spatially depictive) (Fish and Scrivener, 1990).

The designers' relationships with mental models or elements of these models also vary with some being quite fixed or static, and others highly dynamic and kinetic. As earlier described by Piaget (1971), these designers imagined objects of their focus rotating, or their body and viewpoint changing in relation to the object of focus. Through imagination, aided in some cases by digital software, existing forms, shapes and qualities also could be transformed. In addition, designers' responses highlighted various roles for episodic knowledge (direct and vicarious) and contextual knowledge including as sources of inspiration for idea formation through to the constraints and criteria for evaluating successive hypotheses or conjectures.

For some designers, there was heavy reliance on, as well as a strong belief that contextual knowledge such as site conditions, economics and client requirements should be the primary generators of the design. Imagining in this case in the initial stages tended to be textual and descriptive rather than pictorial or spatial. In contrast, other designers opted for a less pragmatic, more aesthetic approach in the initial stages.

For those designers, adopting an aesthetic approach meant starting with an emotive theme. Here we see alignment with the concept of the 'frame' proposed by Dorst (2004) and Paton and Dorst (2011). The emotive theme is highly metaphoric and analogous with the quality of the experience planned for in relation to the building or environment. It also informs, and indeed demands, aesthetic contextual engagement with the site, with the clients, and with the environment as it forms, or more precisely, as it is allowed to form itself.

The data collected from the designers used to inform the categories of imagining just described also included comments about various factors infuencing their ability to imagine, particularly in the early stages of the design project. While designers varied in their need and/or ability to manage their environment in order to facilitate imagining, several designers identified the need to minimise interruptions and other distractions by withdrawing to a more quiet physical environment, or even to a natural environment. There was also mention of the need to prepare oneself for the process of imagining by playing music, or going for a walk. The importance of making time for the imagining process was noted, recognising that in many respects this could not be determined or prescribed. Once 'in the zone' it was considered important to allow the subconscious to take over. One designer described being 'in the zone' as an 'out of body experience' involving transportation to, and an immersion in, the design situation. Some designers describe how they occupy this world in addition to the physical world; in other words, they experience a parallel state of being. This was especially evident when talking about the role of sketching and how it supports the imagined space between the paper and the brain, and associated creative and productive thinking. While some designers are perceptually aware of the physical world beyond the paper, others describe a selective awareness of the paper only. Once an idea crystallises, designers recognise the need to engage more consciously with the external world. This aligns with Folkmann's (2010) notion of focussing and defocussing.

Imagining and presence research

As highlighted in Chapter 3, presence is regarded as a basic human experience encompassing everything from subconscious physiological responses through to higher-level cognitive, and even emotional and bodily awareness, an includes actual physical and social interaction. This rather broad context is encapsulated in Chapter 3 as a framework comprising three main categories: non-mediated presence; mediated presence; technology-mediated presence. Informed by Lombard and Jones (2007), presence experienced when in objective proximity to another person or object is classified as *non-mediated physical presence.* Presence which is generated via a non-technological external element such as a book or an internally generated mental model is labelled as *mediated presence*, and *technology mediated presence* is brought about only through digital technology means. In terms of the latter two states, mediated presence enables a person to perceive themselves to be somewhere other than in the physical world. In the case of technology mediated presence, this is the sense of being in a computer-generated or computer-mediated environment, such as CAVEs or VEs. Maximum presence is understood to exist when the environment is able to support the full intent of the user and when the medium facilitating the imagined or virtual environment is experienced as invisible or transparent.

As previously noted, in design practice there is increasing reliance on digital and virtual technology for the purpose of visualising, simulating and documenting yet to be constructed environments or objects. However, despite significant innovation in digital technology (such as in entertainment or gaming), the use of digital and virtual software by spatial designers in the early stages of design is generally restricted to digital 2D and 3D models of building and interiors, their manipulation and/or movement around and through such spaces.

As noted in Chapter 5, which focuses on designers' experience of designing and imagining in designing, while the designers made reference to this software, they also described experiences of imagining in other ways; in non-technology mediated ways. In presence research, non-technology mediated presence is used to describe human experience in traditionally non-immersive and non-interactive environments such as that elicited through reading books, dreaming or daydreaming. Given the emphasis by the designers on non-technology mediated imagining, the research in design methodology that emphasises the significance of mental modelling, and the research on presence that suggests a stronger connection between presence and non-technology mediated presence, this section will explore possible connections between imagining and presence by focussing on non-technology mediated presence. The following section will then bring together the discussion in this and previous two sections to propose a spatial design imagining (SDI) model.

In terms of non-technology mediated presence, this can have a personal dimension (such as perceiving self or part of self in the context of what is imagined, such as in a daydream). If reading a book, a person may perceive themself to be either 'within' the story as a character, or themself to be in parts of the story. When watching a film a person may also perceive themself to be in the scene as one of the characters, thus feeling and responding with the same emotions or reactions as the actor/s. This form of spatial presence or 'sense of being there' is also referred to in terms of 'physical presence', 'a sense of physical space', 'perceptual immersion', and 'transportation', and may have associated with it a social presence involving the perceived existence of others.

If presence in this sense is regarded as an experience of being somewhere other than in the present physical environment, what facilitates this? Is it the medium alone? Is it something external and perceptual, or internal and conceptualised?

As noted previously, Schubert and Crusius (2002) assert that presence is not a direct result of immersion, but is "mediated by cognitive representations that are constructed on the basis of immersive stimuli" and that it is the "structure of this mental model [that] determines whether the user experiences a sense of presence or not" (Schubert & Crusius, 2002, p. 1). Further to this, they acknowledge cognition as a mediator between immersion and presence. This then begs the question of a possible role of imagining, or as termed in presence literature, imagination.

In expanding on the notion of degrees of absorption in a text, Ryan (2001) identifies one type as 'imaginative involvement', where the reader adopts a "split subject attitude" and "transports herself into the textual world but remains able to contemplate it with aesthetic or epistemological detachment" (Ryan, 2001, p. 10.). Another degree is 'entrancement', described as total immersion where the reader is caught up in the textual world so that the surrounding world fades away and the reader feels as though they are taken to the 'world in the story'. Ryan (2001) also refers to addiction, which is the attitude and the willingness of the reader to escape from reality and be so immersed in the textual or narrative world that they feel able or compelled to interfere with it. When in this state, Ryan (2001) asserts that the individual is most likely to experience spatial, temporal and emotion immersion; transportation to the geography of the narrated world; engagement with the unfolding events of the story; and identification with the characters portrayed.

From the discussion presented in the previous chapters, parallels can be drawn with Ryan's descriptors of degrees of absorption in text with imagining in design. In fact, these cannot be considered discrete concepts, but rather, are similar and integrated notions. For example, it appears from the research presented thus far, designers by the very nature of the context of their creative practice have to adopt a 'split subject attitude', with one dimension facilitating existential engagement, and the other an epistemological and aesthetic detachment facilitating interference with the world; a world where they are both occupant and creator. This suggests that conceptually, certain design activities resemble a strong similarity to presence proposed by Gysbers et al. (2004) that spatial presence so integral to this is facilitated by "higher-order mental activities such as cognitive involvement and imagination" (Gysbers et al., 2004, p. 13).

The discussion in this section has emphasised an internal/conceptual view of presence in an attempt to highlight salient aspects of the relationship between presence and imagining. Likewise the attention now shifts to Biocca's (2003) 'three pole model' and its focus on the role of mental imagery space. According to Biocca (2003), and as described in detail in Chapter 3, the original 'two pole model' of presence only considers virtual and physical spaces, but not imaginary spaces. In contrast, the 'three pole model' allows for the role of mental imagery space. For Biocca (2003), a sense of presence can occur when individuals have the impression that they have 'withdrawn' from physical space into a purely imagined or as he describes it 'imaginal' space. This experience happens when the individual withdraws 'focal attention' from incoming sensory cues, and instead chooses to attend to 'internally generated mental imagery'. This is highlighted in the previous section where designers describe how they facilitate this by withdrawing to a quiet place or a place where there are minimal distractions.

Sensory stimulus is also discussed in presence research in terms of the relationship between presence and imagining. As noted previously, in relation to spatial imagination, Wirth et al. (2007) propose that "spatial imagination becomes more relevant if the mediated representation of the space is less intuitive and more fragmented (e.g., when reading textual descriptions)" (Wirth et al., 2007, p. 502) and this aligns with Wallach et al. (2010), who argue that imagination does not have a significant impact on presence in sensory-rich environments. Jacobson (2002) suggests that imagery abilities may play a role when sensorial cues are less available, and they can, to some extent, stand in for the missing perceptual information (Jacobson, 2002). This is the case with reading books, or as we have noted, with designers in the early stages of designing when information is highly fragmented and vague. Despite the absence of any immediate perceptual stimulation, the designer is able, like the reader (Green, Brock & Kaufman, 2004) to experience a high degree of immersion or a form of presence.

In general, imagination (or in the same sense imagining) involving mental simulation, memory and embodied cognition, is understood by many presence researchers as fundamental to the experience of presence.

The Spatial Design Imagining model

The previous sections have each focussed on imagining in a particular context: first as described in design methodology literature; second, in terms of what emerged from the empirical study conducted as part of this book; and last, as revealed through an exploration of presence research. In this section, these different, albeit complementary, viewpoints are merged to advance a more comprehensive model of imagining of particular relevance to spatial designers and design educators. Figure 6.4 represents the Spatial Design Imagining (SDI) model as presented below.

As highlighted in the overview of design methodology research, the design context is now widely understood as complex, ill-defined, and uncertain. In this regard, all design projects are unique and novel, demanding a heuristic, iterative and satisficing approach. For design educators these qualities pose specific challenges. How, for example, does a design educator support and facilitate student learning for such dynamic and unknowable contexts? The general response has been to engage students in design projects (sometimes real, sometimes hypothetical) with 'expert' tutors who provide feedback informing further action and refinement. Over a period of time and involvement in a variety of projects of varying complexity and typology, students develop a repertoire of design skills and knowledge that enables them to produce 'satisficing' outcomes as judged appropriate in the educational context. How they do this remains largely a mystery for it is only when students externalise their thoughts that a response can be actioned. In the main, the design educators involved in designing and implementing design curricula particularly for design studio learning do so with only tacit understanding of how the students cognitively, affectively and experientially develop their ideas. Indeed, even the students are largely unaware of what they are doing and what is at play especially in the early stages of concept and schematic development. While recent industrial design biased research provides further insight into design thinking this is not informing design education particularly the spatial design areas of architecture and interior design.

We now know for example the major role played by abductive thinking in designing as part of a process that is productive and generative; in other words, creative. We are aware of attempts to further understand this through concepts like imaging, visualisation, mental imaging, imagination, and the like. While these concepts continue to be used interchangeably with no clear definition and thus creating confusion, what is emerging from this study is a growing recognition of the role of what this book refers to as imagining.

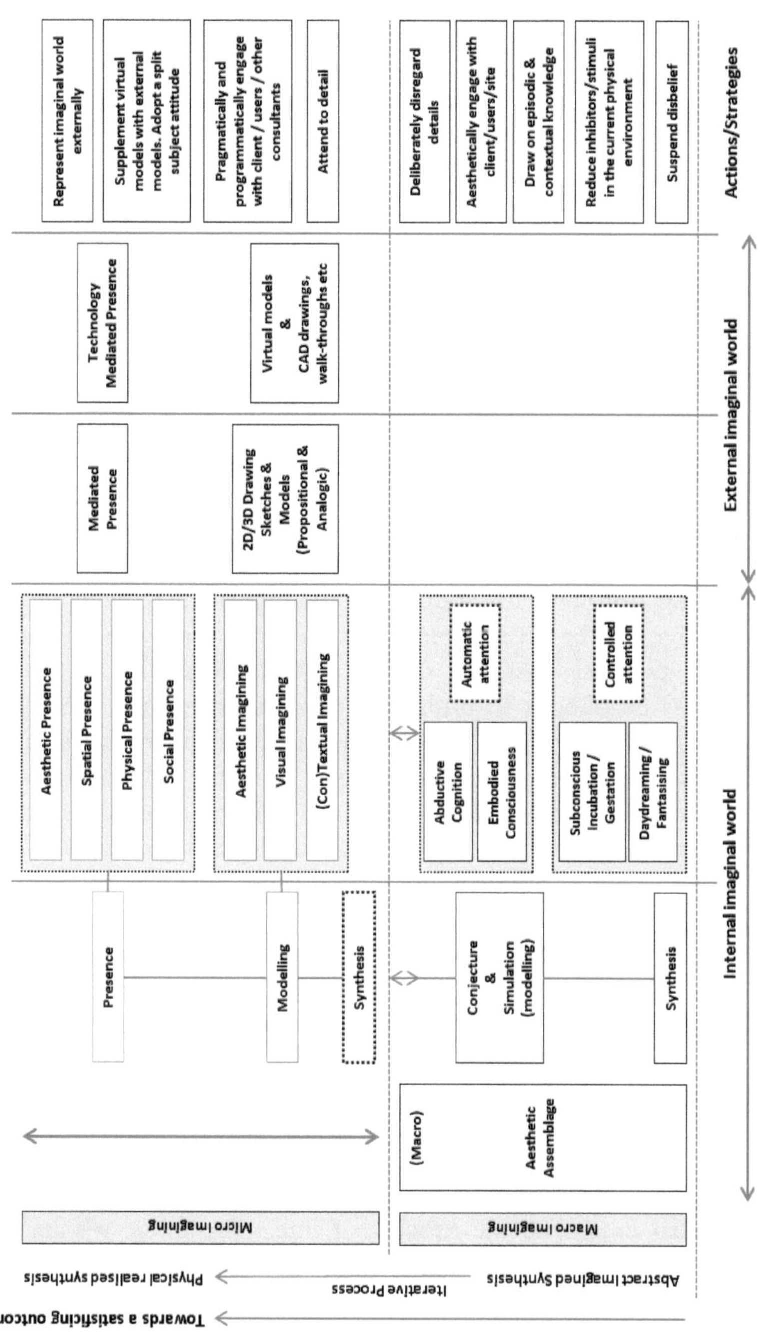

Figure 6.4: The Spatial Design Imagining (SDI) model

What is imagining?

As conveyed diagrammatically in the SDI Model (Figure 6.4), in the context of spatial design by experienced designers, imagining has the following features:

- Synthesis
- Orientation to the future demanding conjecture and simulation, also referred to as modelling
- Simulation that involves imagined transportation to, immersion in, contextual engagement and intervention with the modelled environment.

In the early stages of designing, imagining involves a process of construction, deconstruction, reconstruction, or, as described here, synthesis where an abstract assemblage or aesthetic mental model is generated as a higher order construct or 'frame' for additional more micro level modelling. The process can extend from daydreaming and fantasising through to abductive and speculative thinking. The assemblage can be an idea, concept or theme that, while not totally defined, possesses as part of its metaphoric quality a sense of wholeness enabling it to bridge the problem space and the solution space. In these early stages such a process is internal; that is, the generative process is internalised both consciously and subconsciously. Whilst what is imagined may have visual qualities, in some cases it may be a feeling or a textural concept like 'transition', or an experience encompassing several of these qualities. In this book, visualisation, involving an image produced mentally and seen 'in the mind's eye' such as a symbol or a representation of perceived reality, is referred to as mental imaging. Imagining can involve mental imaging. However, in the generation of an aesthetic mental model that is more defined and represents a future environment it becomes multidimensional; it becomes a 'place' exhibiting affective, existential and temporal qualities as well as physical qualities; a place the designer and others can inhabit and experience.

For spatial designers, the process of imagining these 'wholes', (the process of synthesis) is oriented to the future, to a yet to be constructed and experienced building and/or interior environment. It is in this sense *creative* synthesis. Imagining for this purpose involves speculation and conjecture about what the artefact could be and how it could/should perform and be experienced. The process is heuristic with the designer repeatedly moving between 'parts' and 'wholes'. It is also iterative with ideas thrown out or reworked; a deconstructive and reconstructive process.

The mental models referred to earlier are integral to this creative synthesis process. They provide opportunities and 'spaces' and 'places' for simulation and testing. As can be seen from the designers' responses, they have different types of relationships with these models ranging from detached, static relationships to immersive, engaged relationships involving others or the designer alone. In this study, these models are proposed to comprise the primary categories of (con)textual imagining, visual imagining and aesthetic imagining. Further to this, visual imagining can be pictorial or spatial constituting secondary categories. Deeper analysis of these categories of imagining suggests a relationship between the type of model and the type of simulation. For example, pictorial imagining involving floor plans and elevations appears to restrict simulation to functional and perceptual purposes, while spatial imagining invites and supports imagined bodily immersion at a deeper experiential level.

Depending on the project, stage of designing and/or imagining ability, designers supplement mental models and imagining with external models. These can range from simple hand drawn sketches to sophisticated virtual spaces generated digitally. Substantively and procedurally, designers draw on various types of knowledge, some of which is developed directly from their experiences as well as from research of the context. This includes the experiences of their clients, some of which is vicarious and empathetically imagined.

As described in Chapter 5, the designer's relationship with internal or external models varies. For some, the relationship is detached; for others it is highly engaging, producing a more intense sense of presence. In terms of the latter, this involves imagined transportation to, immersion in, and contextual engagement and intervention with the modelled environment. Presence, as discussed previously, can be either mediated by some form of digital or virtual technology, or non-mediated by other means such daydreaming, reading, sketching, all of which involve some form of mental modelling. The nature of the presence experienced by the designer can be spatial, perceptual, or social.

While it is not highlighted in the presence literature, this book also proposes that presence can be aesthetic in the full existential sense of the word; a form of presence involving embodied consciousness encompassing thoughts, feelings and actualisation. Additionally, the findings illustrate that designers are able to adopt a 'split subject attitude' when in the imagined space. This allows the designer cognitively 'detach themselves' to the point of intervening with, and changing the space or the narrative in response to their experiences of it. In this respect, they are both *occupant* and *creator*, and can be so engaged in this process that the physical world appears to 'fade away'. Both presence literature and the empirical findings lend support to the theory that the intensity of the immersion is related more to an initial lack of information (including sensory information) than to a situation where information is highly detailed.

The degree of sensory information is also relevant to the level of presence experienced. In the study, several designers discussed the necessity to minimise distractions in order to facilitate transportation into the imagined world. This facilitates subconscious gestation or automatic attention, as well as conscious and attentive modelling. Some designers become very expert at being able to straddle the physical world, the imagined world, and in some cases, also a digital virtual world.

Overall, the outcomes of the empirical data and the literature on presence and design research illustrate that imagining effectively supports designers at a macro, highly abstract synthesising level in the initial stages of designing. This assists in the development of an aesthetic mental model that bridges the 'problem space' and the 'solution space', thus facilitating a more refined development towards the solution space and a satisficing outcome. Once this mental model is developed, imagining operates at a micro level and then interchangeably between the macro and micro levels when undertaking simulation and facilitating aesthetic presence.

Also evident directly and indirectly in the designers' responses is the role of attitude in facilitating engagement with the process of imagining. Some designers quite blatantly reject it particularly in the initial stages of designing. This is in contrast to existing theory and the findings of this book that strongly suggests the opposite. In exploiting imagining, designers need to be willing to suspend disbelief and to participate fully in hypothesises that may at first appear unremarkable or highly fanciful at this initial abstract internal level. They need to be willing to vicariously engage with others; bringing others into their imagining and as well as transporting themselves into others' 'being'.

Chapter 7: Concluding the Study: A discussion and conclusion

Introduction

As introduced in **Chapter 1**, this book was prompted by the desire to enhance design learning for students in the spatial design disciplines of architecture and interior design. The role of imagining was of particular interest and how a richer understanding might be developed through an integration of extant theory and empirical research, therein providing the basis for future consideration by design educators. In summary, then, the main research question was:

- What is imagining in the spatial design process?

To address this question, several subsidiary questions were posed, including:

- What is imagining as experienced by interior design and architectural practitioners?
- What extant theory is relevant and, when considered with current empirical research, is likely to expand current knowledge of imagining?
- What are some initial implications for spatial design education and future research?

The purpose of this chapter is to describe and explain how the doctoral project has responded to these questions. It also highlights its limitations and the potential of these limitations, as well as the outcomes of the project providing for future research and consideration. It commences by describing and explaining how our understanding of imagining, and the role of imagining in spatial design has been enriched through this research and then what this proposes for spatial design education and further research. It also highlights additional outcomes and how these contribute to other forms of substantive and procedural knowledge.

Significance and contribution in enhancing knowledge of imagining and its role in spatial design

Procedurally our understanding of imagining, and the role of imagining in spatial design has been enriched by:

- Adopting a Grounded Theory approach (**Chapter 4**) involving empirical research and theoretical saturation through the integration of extant theory with the empirical findings, as well as
- Considering, as part of the extant theory, literature deemed to be relevant (presence theory) but outside the design disciplines.

Substantively, the main outcome is the SDI model that captures, more than before, the extensive and integral role imagining plays in spatial design. While an exploration of design methodology literature summarised in **Chapter 2** produced contextually relevant information for the project, very little was available that explicitly focussed on imagining and that differentiated the cognitive and corporeal activities involved in designing and how they are interconnect at both macro and micro levels of designing. What was in part useful, as highlighted in the SDI model (**Chapter 6**), was recent research on abductive thinking and its role in the process of synthesis integral to novel, highly complex, future-oriented situations. While the emerging research on abductive thinking and framing has extended research relating to creative mental synthesis, it failed to respond convincingly to the call by Kokotovich (2000) to produce a detailed and more transparent understanding sufficient to support design education. It also (inappropriately) assumed relevance across all design disciplines.

As this doctoral research has demonstrated, closer examination of how spatial designers undertake design and the role of immersion in their designing significantly extends our understanding of designing generally as well as more specifically in the interior design and architecture disciplines. This was described in detail in **Chapter 5**, which presented the results of an empirical study of spatial design practitioners' designing and imagining experiences. The outcome is encapsulated in the taxonomy of imagining qualifying imagining as visual (pictorial and spatial), (con)textual, and aesthetic. Emerging from this research is greater recognition of the role of mental and external modelling and their interrelationships, of controlled attention in the form of embodied consciousness, as well as the subconscious in terms of incubation and gestation or what may be termed automatic attention. Identification of aesthetic imagining emphasising embodied consciousness is particularly interesting and is explored further below in terms of potential implications for design educators. The empirical study also highlighted the role of the physical environment for imagining, and actions, strategies and attitudes employed by designers in managing complexity and uncertainty aesthetically.

Integrating the outcome of a review of presence research (**Chapter 3**) added significantly to the emerging theory in this doctoral research. In particular it extended an understanding of how one engages with complexity through immersion and how this can be mediated through technology or otherwise. Of particular note is how less artificial mediation and a lack of detail and stimuli appears to support more aesthetic engagement. The review of presence literature identified that presence is generally restricted to spatial, physical and social forms of presence, and as such, the inclusion of aesthetic engagement was not found in current literature. In this way, the empirical study makes a contribution to presence research. In terms of presence research, it did however also highlight the role of imagining as a mediator between immersion and presence with the taxonomy of imagining developed from the empirical study also informing how this is, and can be, operationalised by designers. Presence research also contributed through its discussion on the internal/conceptual view versus the external/perceptual view. While presence research regards these as discrete, the empirical study suggests otherwise. For example, as expressed earlier, it appears designers by the very nature of the context of their creative practice have to adopt a split subject attitude, with one dimension facilitating existential engagement and the other an epistemological and aesthetic detachment facilitating interference with the world; a world where (as largely disregarded in presence literature) the participant is both occupant and creator.

In all, the doctoral research responds to a neglect in design research literature of the role of imagining in spatial design by recognising designers' experience of imagining and the role of presence research in revealing new knowledge and a new framework for understanding imagining. Specifically, it has unpacked the phenomenon of imagining, identifying and providing further clarity of the elements and their relationships. In doing this it has produced an enriched basis for translation to other design disciplines as well as to design education, particularly spatial design education, which is explored further in the following section along with its implications for further research.

Implications for spatial design education specifically design studio teaching

This section explores the implications of the findings for spatial design education. It does this through a focus on the design studio, the main vehicle used by design educators to teach design process.

The exploration of imagining undertaken in this book has revealed the relationship between the internal imaginal world and the external 'actual' world as central to better understanding the role of imagining and enriching design process and outcomes. It has also revealed that designers are influenced by a range of factors and that they adopt particular actions and strategies when negotiating the internal world, the external world, and the interstice between.

The empirical data illustrated that designing and the design context do not sit in isolation. Rather, the design 'problem' and 'solution' sit within broader and more complex relationships concerning the designer, the users of the future space, and the client. Additionally, the design situation also sits within a context and, as such, will always impact in some way or another, the broader community. Teal (2010) supports this, describing how the problem or design context, "has far-reaching implications on such interconnected groups and phenomena as family, community, culture, class, economy and employment" and that designers should work in a manner that requires "immersion and involvement" (Teal, 2010, p. 5) with these factors and relevant communities. For Teal (2010), existing culturally engrained practices of design actively work against this, limiting the nature and extent to which this occurs; in the process neglecting to address and respond to the "the complexities, accidents and flows that that are basic to a dynamic and vital existence" (Teal, 2010, p. 2).

As he argues, traditional design studio education generally follows a rationalistic linear path, being somewhat codified to the standard stages of professional design practice such as 'brief development', 'concept development', 'schematic design' and 'design development', 'construction documentation' and 'construction administration'. For students, any contact with the design context is for the most part hypothetical and confined to the immediate client and user through the brief development stage. Where students do engage with other people as real or simulated clients and users, any additional contact is normally restricted to feedback in the schematic and design development stages. The higher education community with its systems and processes, further impact on this situation. Increasing class sizes, restrictive time slots and spaces for learning, assessment processes, and risk minimisation policies all conspire to disengage the student designer from the design context. In all, the potency of design in education and practice is severely compromised; a situation further exacerbated by a lack of knowledge of the embodied cognitive nature of design and its role in managing and capitalising on the creative richness of ambiguity.

As the empirical study shows, while there are designers who place initial emphasis on the external physical and functional requirements and eschew as folly any consideration of affective or existential aspects, there are others who commence by adopting an aesthetic, internalised process; embracing embodied, intuitive and emotional knowledge and fluidity of practice at the beginning as well as throughout the process. While these designers implicitly realise the need for, and the benefits of this approach, there has been no research that adequately qualifies this. By considering extant theory from presence research this book provides a deeper understanding for spatial design and the significance in adopting this approach. For example, it appears that less (rather than more) pragmatic detail inspires and informs the generation of an increased number of ideas, that the generation of these aesthetic wholes early on are central to producing innovative, robust outcomes, and that the abductive development of these wholes or aesthetic assemblages is facilitated more effectively when undertaken initially and primarily in the internal imaginal world of the designer. In contrast, design studio teaching, as is the case with practice, places emphasis on collecting as much pragmatic detail as possible early on and building towards a whole over the course of a number of stages involving as early as possible the externalisation of ideas through sketching and drawing and movement between the internal imaginal world and the external world. As Teal (2010) points out, invariably this leads to inertia as students "fish around for the *best* idea before moving forward" (p. 3). In this respect, "there is no right way to proceed except not to proceed; everything is connected to everything else" (Teal, 2010, p. 4).

The discussion in this section has emphasised the conceptual work of Teal (2010) and drawn connections between his work and the findings of the book that relate to the aesthetic whole. This is because the notion of the 'rhizome', which he borrows from Gilles Deleuze and Felix Guattari, has strong parallels with the aesthetic whole or assemblage as it is described in this book. For Teal (2010), the visualisation of the rhizome suggests immersion and dynamic experimentation, operating in a rhizomic way within, rather than reductively from without; it depends on one doing less analysis and more production. As he highlights in line with Deleuze and Guattari, and which has emerged in this research, the initial whole, assemblage or design concept is whole because it totalises its components, but it is a fragmentary whole remaining ready to respond and change. In this respect, acting rhizomically for a designer or design student means inhabiting an emerging solution in a process of active participation and constructive exploration; an interpretive process underlined by an "anticipatory projection of meaning" (Snodgrass and Coyne, 2006, p. 38). The findings of this book including its emphasis on imagining provide further information on how this may be facilitated in the design studio.

One of the major responses in advocating this holistic aesthetic approach will relate to the assertion that students are not experienced designers. Some are quite young, and as such, do not have diverse and substantial episodic and contextual knowledge to draw upon. However, irrespective of age and design experience, students are embodied beings. They have a phenomenological sense of a being within the environment moving around, seeing, smelling, touching and hearing. This book invites educators to consider such experience a rich source for aesthetic imagining at both macro and micro levels, in the process enabling the student to develop a greater awareness of self in relation to 'other'. It also invites explicit focus on the nature of abductive thinking and how to frame situations through greater consideration of the role of the built environment in the broader social, political and environmental context, in addition to a detailed exploration of the client and user needs and desires, both espoused and tacit.

Emerging research by interior design educators and researchers such as Poldma (2009) highlights the need for two inter-related concepts to be explicitly acknowledged in the design process: the dynamic inter-relationship between the designer and the social reality of the designed space; and the co-construction of the problem and solution space with users and stakeholders (Poldma, 2009, p. 4671):

> Using these processes, the interior designer uncovers the existing social and political framing the problem, observes and discusses needs and desires with different users and stakeholders, and through aesthetic design proposals proposes new and alternative solutions that socially and politically democratize and empower those who have experiences in the spaces (Poldma, 2009, pp. 71-72).

This book supports the assertion further proposing the need to focus on this at the commencement of a spatial design course and throughout. In many respects, it constitutes a threshold concept with celebrates and exploits design as interpretation (Rodrigo, 2010). As recognised by Rodrigo (2010) in relation to the work of Meyer and Land (2002):

> [T]hreshold concepts in learning and teaching represent a transformative and irreversible way of understanding a subject, likened to a portal through which a previously inaccessible way of thinking is opened up to the learner and without which the learner cannot progress (Rodrigo, 2010, p. 1).

Traditional design studio education typically requires students to begin conceptual design expressing it as an external process through sketching and drawing from week one of the semester. However, in proposing an alternative pedagogy based on this book, students might be permitted a period of time contemplating and allowing ideas to gestate prior to representing their ideas visually. This places a greater emphasis on the imaginal space being liminal territory where creative ideas (which are as yet unformed) percolate with 'inhabitation' of the problem space and the solution space allowing for (directed and purposeful) fantasising and daydreaming, and embodied consciousness. The suspension of disbelief, which is an essential component of imaginal space, enables the students to incubate information without being too constrained or being under pressure to produce and present ideas before they understand the context or design problem in its abstracted complexity. This initial period also welcomes closer engagement with other people and understanding their lived experience of the world. Rather than going away and working alone on developing a concept, the book invites educators to emphasise dialogue and the development of a narrative as a space for immersion and exploration: "when designers and users construct their experiences together, then tacit forms of knowledge emerge through the experiences that are considered during various stages of the design process. This tacit nature of the design process becomes part of the experiential knowledge that is vital to understand how people appropriate and use interior spaces for their daily lives" (Poldma, 2009, p. 73).

Some key pedagogical components emerging from this discussion include:

- Intuition: where students generate immediate associations to the design task through their own personal experience of the world and what they know about designing through their design education.
- Sensibility: where students are able to evoke a form of response or feelings during the cognitive design process that involves a broader aesthetic view; preferably one that is co-created in a dialogic situation with others.
- Productivity: where students productively generate ideas – even ideas that initially appear flawed.
- Exploration: where students, with others, explore the unknown through mental modelling and aesthetic immersion.
- Novelty: where students are encouraged to create uncommon concepts and explore novel ideas.
- Attention: where students formalise ideas through automatic and controlled attention.
- Elucidation: where students express abstract ideas experimenting initially with narrative forms of communication.
- Assessment: where students are recognised and rewarded for taking risks.

These approaches also invite discussion about assessment. Typically educators assess students on what they produce. This may involve:

- Project proposal: outline of a proposed project detailing aims and objectives, materials and methods, purpose and context, performance criteria etc.
- Models: conceptual models, work-in-progress models and individual prototypes
- Drawings/sketches: representational media including conceptual sketches, work-in-progress and final drawings
- Presentation/ 'crit': verbal presentation of design work to an audience of instructors, experts and/or peers
- Portfolio: an organised (curated or edited) collection of a student's work designed to represent their achievements and effort over a period of time
- Reflective journal: a diary that encourages introspective and self-directed learning developed over time.

As conveyed, these include for the most part an emphasis on externalised representations of the end product, interior or building. A possible alternative is that for the first semester assessment focuses exclusively on productive and generative thinking (and being) in relation to a design 'problem' and its aesthetic and existential context. An aspect of this could involve students in critically examining their own cognitive and emotive processes and responses, and exploring various ways in which these can be communicated and shared. Once this initial emphasis has been established, students could then progress to using the more traditional process of moving between internal and external imagining with explicit attention in this case to 'split subject attitude', that is, learning to effectively move between an immersed embodied state and a detached critical state. At the more micro stage, students can be introduced to the different types of imagining (visual and (con)textual as well as aesthetic) and to the techniques and mechanisms that facilitate them. What should become obvious is how certain conventions and practices such axonometric drawing influence the designer's orientation and degree of immersion in the model. Of particular note here is the use of technology in relation to its role in mediating imagining.

The current advances in digital technology have impacted on the design process and practice in ways that even half a century ago, would have seemed impossible. These days, students can access a variety of digital programs (such as Sketch Up) that are free and easy to use to develop drawings of buildings, interior spaces, and so on. Some software products even allow simulated walk-throughs of spaces. There are also technologies (such as rapid prototyping and laser cutting) used by designers, particularly in the conceptual stages of the design process to create complex three-dimensional models. However, findings of this book suggest that these may be a distraction inviting premature emphasis on form and visual qualities at the expense of deeper aesthetic imagining best developed internally in the initial stages. Having said this, the findings of the book do invite further cross-disciplinary development of these technologies to externalise (when appropriate) internal imagining in richer more embodied ways, as discussed previously. Greater provision for, and encouragement of co-presence and collaborative immersion and exploration will help strengthen student/client relationships as well student/educator dialogue, and create a shared understanding about the process of design, not just the tangible/formal outcome of design. This will enable more explicit focussed (as opposed to tacit) development of process-based strategies, as well as increased understanding of the nature of, and the relationship between design research, design theory, design practice, design education and pedagogy.

The discussion to date also suggests reconsideration of studio culture and environment. As highlighted in this research, immersion in and exploration of the internal imaginal problem/solution space demands, for most people, minimisation of external stimuli and choice in relation to where and how they do this. Responding to this may challenge existing practices including the relationship of educators with students as well as the provision of resources and facilities. For the most part, 'the studio' is currently chiefly a physical space on campus where studio project based learning occurs within a designated time in the presence of fellow students and tutors. This book suggests greater flexibility for students to choose where, when and with whom to undertake the initial stages of the project. The potential of technology to enhance imagining and engagement adds another dimension to 'studio learning'. The book proposes that the current initiatives regarding technology and blended learning also include consideration of the main points raised by this book and highlighted in this chapter. These emerging developments provide fertile ground for expanding the outcomes of this research, and incorporating areas that have been excluded from this study such as the role of communication in imagining.

Understandably, this will not be without challenges given decreasing resources, an increasing risk-adverse higher education environment and university demand for quality assurance involving greater transparency and learning outcomes linked directly to unit aims and objectives. As such, engaging with the ideas presented in this bookwill demand a holistic approach involving curricula and associated higher education processes and practices as well as professional accrediting and registering body requirements.

Wider implications and contribution

To a large extent design cognition has remained unchartered territory. Perpetuating this is the belief that design ability stems from creative talent – it is mysterious and inaccessible; something that many practitioners exploit for various reasons in their relationship with clients. However, as Polaine (2011) points out: "We have sold what we do as magic at the cost of hiding our processes, and when we hide our processes we can no longer articulate them, teach them or give them the value they deserve" (Polaine, 2011, p. 44). In this respect, the potential of design to help address complex social and environmental issues is severely restricted. Exacerbating this is the increasing popularity of STEM – science, technology, engineering and mathematics. In all, "design research has failed to ignite public imagination with the rhetoric of STEM dominating the media" (Polaine 2011, p. 41) and, as I have witnessed in my own university, the distribution of university teaching and research resources.

As revealed in the literature review, attempts to value add through design have chiefly come from the industrial design discipline and the work of Dorst (2011) and others in exploring how design thinking can be appropriated by other disciplines such as business or marketing. Such popularisation of design thinking (that is, the notion that non-designers can think like a designer) has led only to further estrange the strategy and process of design from embodied experience and sensation (Tonkinwise 2011; Stewart 2011). According to Stewart

(2011), the focus needs to shift from production and functionality to cognition, experience, processes, interfaces and relationships (Stewart, 2011, p.1).

This book has addressed these concerns in three ways: first, it has focussed on developing greater clarity in the early stages of design, stages that have traditionally been described as mysterious and magical; second, it has attempted to do this in a holistic way capturing the aesthetic as well as pragmatic aspects of the process; and third, it has undertaken the research from a spatial as opposed to a product perspective.

As outlined, the findings of the book have also been informed by extant theory outside the design disciplines, namely that developed in relation to presence research. There are many reasons to unify fragmented spheres of knowledge with concepts and methods in one discipline assisting in identification, understanding problems and issues in another (Aram, 2004, p. 382). In this book, presence research has been instrumental in extending our understanding of the role of imagining in immersion and creative synthesis. Equally, presence research has been informed by the empirical research conducted through this book and its elucidation of the aesthetic quality of imagining and the 'split subject attitude' of the designer. A deep conceptual understanding of the process of imagining in design in these respects is noticeably absent in presence literature. Given the findings of this research which highlight the complimentary relationship between imagining and presence, a bridge has been built between the two domains establishing grounds for future dialogue, exploration and 'epistemological interdisciplinarity' (Aram, 2004).

Conclusion

As outlined, this book was prompted by a personal interest in more effectively supporting spatial design students in the early idea generation and formation stages of designing. Responding to this and the associated need to better understand imagining in the spatial design process, a multifaceted research project was undertaken seeking to address the following sub-questions:

- What is imagining as experienced by interior design and architectural practitioners?

- What extant theory is relevant and, when considered with current empirical research, is likely to expand current knowledge of imagining?
- What are some initial implications for spatial design education and future research?

By bringing together extant theory from design methodology research, the empirical study findings, and with extant theory from presence research, two outcomes emerged that make a foundational contribution to addressing the first two questions. These are the categories of imagining taxonomy (Chapter 5) and the spatial design imagining (SDI) model (Chapter 6), which emerged by employing a Grounded Theory informed study. From within the context of Grounded Theory methodology, the SDI model is an 'illustrative model' of 'substantive Grounded Theory' possessing both persuasive explanatory power as well as the potential for application (Birks & Mills, 2011). This is demonstrated in Chapters 5 and 6 as well as through the discussion in this chapter exploring some initial implications for spatial design, its education, and future research. It also stands in relation to its potential to bring about practice change and develop new knowledge in and for the discipline.

While the theory generated through this research may not be applied in its totality due to limitations of sampling size and geographic context, it has produced a strong theoretical foundation for future research involving the development of a formal Grounded Theory that has applicability across a number of substantive areas, and the development of strategies for implementation aimed at practical change in spatial design education.

In this respect, the book makes the following recommendations:

- Further research involving a wider participant pool internationally as well as nationally within the spatial design discipline to strengthen the theory of spatial design imagining and spatial design process
- Further empirical research across design disciplines (architecture, interior design, landscape architecture, industrial design, fashion and interactive design) to consolidate its discipline as well as multidiscipline relevance

- Further research outside the design disciplines such as in the presence domain to extend existing extant theory relevant for imagining and creative mental synthesis and develop opportunities for interdisciplinary research
- Development of implementation strategies by spatial design educators to apply, evaluate and extend the SDI model's value in spatial design education
- Experimentation with various design and research methods that complement Grounded Theory or an alternative 'meta' methodology.

In conclusion, these recommendations, building on the foundational research of imagining described in this book will contribute to a richer experiential understanding of creative mental synthesis. While the "the skills of synthesis, of making connections between disparate fields and data points, of making intuitive leaps based on past experiences and insight are crucial to dealing with a world that is in constant flux and whose rate of continuous change is only going to increase…" (Polaine, 2011 citing Johnson, 2010, p.43), further to this is an embodied consciousness and ability to inhabit a problem space and an emerging solution so as to respond with meaningful complexity rather than superficial affectivity or levelled uniformity (Teal, 2010, p. 9).

References

Akin, Ö. (1979/1984). An exploration of the design process. In N. Cross (Ed.), *Developments in design methodology* (pp. 189-207). Chichester, England: Wiley (Originally published in *Design Methods and Theories*, 1979, *13* (3/4), 115-119).

Akin, Ö. (1986a). A formalism for problem restructuring and resolution in design. *Environment and Planning B: Planning and Design, 13*, 223-232.

Akin, Ö. (2001). Variants in design cognition. In C. Eastman, M. McCracken & W. Newstetter (Eds.), *Design knowing and learning: Cognition in design education* (pp. 105-124). Amsterdam: Elsevier Science.

Albaum, G. (1997). The Likert scale revisited: An alternate version. *Journal of the Market Research Society, 39*, 331-349.

Alexander, C. (1964). Notes on the Synthesisof Form. Cambridge, MA: Harvard

University Press.

Amaratunga, D., Baldrey, D., & Sarshar, M. (2001). Process improvement through performance measurement: the balanced scorecard methodology. *Work Study* 50 (4/5) (2001), pp. 179-188.

Annells, M. (1996). Grounded theory method: Philosophical perspectives, paradigm of inquiry and postmodernism. *Qualitative Health Research*, 6(3), pp. 379-393.

Annells, M. (1997a). Grounded theory method, part I: Within the five moments of qualitative research. *Nursing Inquiry*, *4*(2), 120-129.

Annells, M. (1997b). Grounded theory method, part II: Options for users of the method. *Nursing Inquiry*, *4*(3), 176-180.

Annells, M. (1997c). The impact of flatus upon the nursed. Unpublished doctoral dissertation, Flinders University of South Australia, Adelaide.

Appleton, J. (1997). Constructivism: A naturalistic methodology for nursing inquiry. *Advances in Nursing Science*, *20*(2), 13-22.

Aram, J. D. (2004). Concepts of Interdisciplinarity: Configurations of Knowledge and Action. *Human Relations*, 57 (4), 379-412.

Archer, L. (1965). Systematic Method for Designers. London: The Design Council

Archer, L. (1980). A View of the Nature of Design Research, *Design, Science, Method: proceedings of the 1980 Design Research Society Conference*, (Jacques, R., & Powell, J. A. Eds.), UK: Westbury House.

Asimow, M (1962). Introduction to Design, Englewood Cliffs, NJ: Prentice-Hall, Inc.

Arnheim, R. (1969). Visual thinking. Berkeley: University of California Press.

Athavankar. U. A. (1996). Mental Imagery as a Design Tool. *Cybernetics and Systems*, 28 (1), 25-42.

Athavankar, U. A. (1997). Learning from the way Designers Model Shapes in their Mind, Cognitive Systems: From Intelligent Systems to Artificial life? ed. J.R. Issac and V. Jindal, Tata McGraw-Hill, New Delhi, 1997, pp 221-232.

Athavankar (1999) Gestures, Imagery and Spatial Reasoning, 'Visual and Spatial Reasoning', Eds.John S, Garo and Barbara Tversky, Preprints of the International Conference on Visual Reasoning (VR 99), MIT, June 15-17, 1999. pp 103-128.

Auerbach, C., & Silverstein, L. (2003). Qualitative data: An introduction to coding and analysis. New York: NYU Press.

Bachelard G. (1971). (trans. 1965). The Poetics of Reverie. Boston, Beacon Press.

Bailenson, J. N., Blascovich, J., Beall, A. C., & Loomis, J. M. (2003). Interpersonal distance in immersive virtual environments. *Personality and Social Psychology Bulletin*, 29, 1-15.

Ball, L. J., & Christensen, B. (2007). *Analogical reasoning and mental simulation in design: Two strategies linked to undertainty resolution.* Paper presented at the International workshop on Design Meeting Protocols (DTRS7, the 7th Design Thinking Research Symposium), London, September 18-21.

Ball, L .J., Evans, J. S. B. T., & Dennis, I. (1994). Cognitive processes in engineering design: A longitudinal study. *Ergonomics, 37*(11), 1753-1786.

Ball, L. J., Ormerod, T. C., & Morley, N. J. (2004). Spontaneous analogising in engineering design: A comparative analysis of experts and novices. *Design Studies, 25*, 495-508.

Banos, R. M., Botella, C., Guerrero, B., Liano, V., Alcaniz, M., & Rey, B. (2005). The third pole of the sense of presence: Comparing virtual and imagery spaces. *PsychNology Journal*, 3, 1, 90-100.

Bayazit, N. (2004). Investigating Design: A Review of Forty Years of Design Research. *Design Issues*: 20 (1).

Beer, R. D. (1995). A dynamical systems perspective on agent-environment interaction. *Artificial Intelligence, 72*, 173-215.

Bernstein, R. (1983). Beyond objectivism and relativism: Science, hermeneutics, and praxis. Philadelphia: University of Pennsylvania Press.

Bhatta, S. R., & Goel, A. K. (1997). An analogical theory of creativity in design. In D. B. Leake & E. Plaza (Eds.), *Case-based reasoning research and development. Proceedings of ICCBR'97, the 2nd International conference on case-based reasoning, Providence, RI, USA, July 25-27, 1997* (pp. 565-574). Berlin: Springer.

Biocca, F. (1997). 'Is This Body Really "Me"? Self Presence, Body Schema, Self-consciousness, and Identity' in The Cyborg's Dilemma: Progressive Embodiment in Virtual Environments, JCMC (3)2.

Biocca, F. (2002). Presence working group research targets. Presentation at the 'Presence Info Day' of the European Commission, January 10, 2002, Brussels.

Biocca, F. (2003). Can we resolve the book, the physical reality, and the dream state problems? From the two-pole to a three-pole model of shifts in presence. Presented at the EU *Future and Emerging Technologies Presence Initiative Meeting.* Venice, May 5-7, 2003.

Biocca, F., Harms, C., & Burgoon, J. (2003). Towards a more robust theory and measure of social presence: Review and suggested criteria. *Presence: Teleoperators and Virtual Environments.*

Biocca, F., & Levy, M.R. (1995). Communication in the age of virtual reality. Hillsdale, NJ: Earlbaum.

Birkerts, S. (1994). The Gutenberg Elegies: The Fate of Reading in an Electronic Age. Boston, MA: Faber & Faber.

Birks, M. & Mills, J. (2011). *Grounded Theory: a practical guide.* Sage Publications, London.

Blackler, D. (2007) Reading W. G. Sebald: adventure and disobedience. London: Camden House.

Block, N. (Ed.) (1981). Imagery. Cambridge, MA: MIT Press.

Blumer, H. (1969). Symbolic interactionism: perspective and method. Englewoods Cliff, NJ: Prentice Hall.

Bogdan, R. & Biklen, S. (1992). Qualitative research for education: an introduction to theory and methods, 2nd ed. Boston, MA: Allyn & Bacon.

Boland, R., & Collopy, F. (2004). Design matters for management. In R. Boland, R. & F. Collopy (Eds.), *Managing as designing* (pp. 3-18). Stanford, CA: Stanford University Press.

Braun, V., & Clarke, V. (2006). Using thematic analysis in psychology. *Qualitative Research in Psychology*, 3: 77-101.

Brown, E., & Cairns, P. (2004). A grounded investigation of game immersion, CHI '04 extended abstracts on Human factors in computing systems, April 24-29, 2004, Vienna, Austria.

Brown, T. (2008). Design thinking. Harvard Business Review, June 2008, 84-92.

Bryant, A., & Charmaz, K. (2007). (Eds.). Sage handbook of grounded theory. London: Sage.

Buck, R. (1999). The biological affects: A typology. *Psychological Review, 106*(2), 301-328.

Buckingham & Shum, S. (1997, March 24-26). *Representing hard-to-formalise, contextualised, multidisciplinary, organisational knowledge.* Paper presented at the AAAI Spring Symposium on Artificial Intelligence in Knowledge Management, Stanford University, Palo Alto, CA.

Burch, N. (1979). To the distant observer. Berkeley: University of California Press.

Burch, N. (1990). Life to those shadows. Berkeley: University of California Press.

Carassa, A., Morganti, F., & Tirassa, M. (2004). Movement, action, and situation: Presence in Virtual Environments. *Proceedings of the 7th Annual International Workshop on Presence (Presence 2004)*, Valencia, Spain, (13-15 October 2004), M. Alcañiz Raya & B. Rey Solaz (Eds.).

Carroll, J. M., & Rosson, M. B. (1985). Usability specifications as a tool in iterative development. In H. R. Hartson (Ed.), *Advances in human-computer interaction* (Vol. 1, pp.,1-28). Norwood, NJ: Ablex.

Charmaz, K. (1983). Loss of self: A fundamental form of suffering in the chronically ill, *Sociology of Health and Illness* 5 (2) (1983), pp. 168-197.

Charmaz, K. (1990). Discovering chronic illness: Using grounded theory. Social Science and Medicine, 30, pp. 1161-1172.

Charmaz, K. (1994). Discoveries of self in illness. In M. L. Dietz, R. Prus, & W. Shaffir (Eds.), Doing everyday life: Ethnography as human lived experience (pp. 226-242). Mississauga, Ontario, Canada: Copp Clark, Longman.

Charmaz, K. (1995b). Grounded theory. In J. Smith, R. Harré, & L. Langenhove (Eds.), Rethinking methods in psychology (pp. 27-65). London: Sage.

Charmaz, K. (1999). From the 'sick role' to stories of the self: Understanding the self in illness. In R. D. Ashmore & R. A. Contrada (Eds.), Self and identity, Vol. 2: Interdisciplinary explorations in physical health (pp. 209-239). New York: Oxford University Press.

Charmaz, K. (2000). Grounded theory: Objectivist and constructivist methods. In Denzin, N. K. & Lincoln, Y. S. (Eds.), Handbook of qualitative research (2nd ed., pp. 509-35). Thousand Oaks, CA: Sage Publications.

Charmaz, K. (2006). Constructing grounded theory: a practical guide through qualitative analysis, London, Sage.

Charmaz, K., & Mitchell, R. (1996). The myth of silent authorship: Self, substance, and style in ethnographic writing. *Symbolic Interaction, 19*(4), 285-302.

Clark, A.E. (1997a). The dynamical challenge. *Cognitive Science, 21* (4), 461–481.

Clark, A.E. (1997b) Being There: Putting Brain, Body, and World Together Again. Cambridge: CUP.

Clark, A.E. (1998). Embodied, situated, and distributed cognition. In W. Bechtel & G. Graham (Eds.), *A Companion to Cognitive Science.* (pp. 506-517). Malden, MA: Blackwell Publishers, Inc.

Clarke, A. E. (2005). Situational Analysis. London: Sage Publication.

Clark, R. E. (2010). Cognitive and neuroscience research on learning and instruction: Recent insights about the impact of non-conscious knowledge on problem solving, higher order thinking skills and interactive cyber-learning environments. Presentation made at the *International Conference on Education Research (ICER)*, Seoul, South Korea, 2010.

Cohen, L., Manion, L., & Morrison, K., (2000). Research Methods in Education,(5th ed.). London: Routledge Falmer.

Coleridge, S. T. (1847). Biographia Literaria, Volume II. London: William Pickering.

Concise Oxford Dictionary (6th edition, 2006). Sykes, J.B. (editor) (Oxford: Oxford University Press).

Conklin, J. (2006). Wicked problems and social complexity. In J. Conklin (Ed.), *Dialogue mapping: Building shared understanding of wicked problems (Chapter 1).* New York: Wiley.

Conrad, J. (1951). The Nigger of the Narcissus: A Tale of the Sea. New York, NY: Harper.

Cook, C. M., & Persinger, M. A. (1997). Experimental induction of 'sensed presence' in normal subjects and an exceptional subject. *Perceptual and Motor Skills,* (85) 683-693.

Corbet-Owen, C., & Kruger, L. (2001). The health system and emotional care: Validating the many meanings of spontaneous pregnancy loss. *Families, Systems & Health, 19*(4), 411-426.

Corbin, J., & Strauss, A. (2008). Basics of Qualitative Research: Techniques and Procedures for Developing Grounded Theory, (3rd ed.). London: Sage Publications.

Craik, K. (1943). The Nature of Explanation. Cambridge: Cambridge University Press.

Creswell, J. W. (1998). Qualitative inquiry and research design: Choosing among five traditions. Thousand Oaks, CA: Sage.

Cross, N. (1982). Designerly ways of knowing. *Design Studies, 3*(4), 221-227.

Cross, N. (1984). Introduction to Part One: The management of design process. In N. Cross (Ed.), *Developments in design methodology* (pp. 1-7). Chichester, England: Wiley.

Cross, N. (2001a). Design cognition: Results from protocol and other empirical studies of design activity. In C. Eastman, W. M. McCracken & W. C. Newstetter (Eds.), *Design knowing and learning: Cognition in design education* (pp. 79-103). Amsterdam: Elsevier.

Cross, N. (2004b). Expertise in design: An overview. *Design Studies, 25*(5), 427-442.

Csikszentmihalyi, M. (1990). Flow: The Psychology of Optimal Experience. New York: Harper & Row.

Damasio, A. (1994). Descartes' Error: Emotion, Reason and the Human Brain. New York: Grosset.

Damasio, A. (1999). The Feeling of What Happens. London: Random House.

Damasio, A. R., & Damasio, H. (1996). Making Images and Creating Subjectivity. In R. Llinás & P.S. Churchland (Eds.). *The Mind-Brain Continuum: Sensory Processes* (pp. 19-27). Cambridge, MA: MIT Press.

Darke, J. (1979/1984). The primary generator and the design process. In N. Cross (Ed.). *Developments in design methodology* (pp. 175-188). Chichester, England: Wiley (Originally published in *Design Studies*, 1979, *1* (1), 36-44).

Dasgupta, S. (1989). The structure of design processes. *Advances in Computers, 28*, 1-67.de Laine, M. (1997). Ethnography: Theory and applications in health research. Sydney, Australia: Maclennan and Petty.

Demian, P. & Fruchter, R. (2005). Measuring Relevance in Support of Design Reuse from Archives of Building Product Models. ASCE *Journal of Computing in Civil Engineering*, 19(2), 119-136.

Dennett, D. C. (1969). *Content and Consciousness*. London: Routledge & Kegan Paul.

Dennett, D. C. (1978). *Brainstorms*. Montgomery, VT: Bradford Books.

Dennett, D. C. (1981). *Two approaches to mental imagery*. In N. Block (Ed.), Imagery. Cambridge, Mass.: The MIT Press/Bradford Books.

Dennett, D. C. (1991). *Consciousness Explained*. Boston, MA: Little, Brown.

Dennett, D. C. (2002). Does Your Brain Use the Images in It, and If So, How? *Behavioral and Brain Sciences* (25) 189-190.

Denzin, N. K., & Lincoln, Y. S. (2000). Introduction: the discipline and practice of qualitative research. In N. K. Denzin, & Y. S. Lincoln, (Eds.), *Handbook of Qualitative Research*, Sage, London, pp. 1-28.

Di Blas, N., Gobbo. E., & Paolini, P. 3D Worlds and Cultural Heritage: Realism vs. Virtual Presence. (2005) In J. Trant & D. Bearman (Eds.). *Museums and the Web 2005: Proceedings*, Toronto: Archives & Museum Informatics, published March 31, 2005

Dimmock, C., & O'Donoghue, T. (1997). Innovative school principals and restructuring life history portraits of successful managers of change. London: Routledge.

Dodson, L., & Dickert, J. (2004). Girls' family labor in low-income households: A decade of qualitative research. *Journal of Marriage and Family*, *66*, 318-332.

Dorst, K. (2004). The Problem of Design Problems - Problem Solving and Design Expertise. *Journal of Design Research* 4(2).

Dorst, K. (2011). The Core Of Design Thinking And Its Application. *Design Studies 32* (6), 521-532.

Durand, G. (1993). The Implication of the Imaginary And Societies. *Sociologie-Contemporaine*, 41(2), 17-32.

Durlach, N. I. & Slater, M. (Eds.) (1992). *Presence: Teleoperators & Virtual Environments*. Cambridge, MA: The MIT Press.

Dunne, D., & Martin, R. (2006). Design thinking and how it will change management education: An interview and discussion. *Academy of Management Learning & Education*, *5*(4), 512–523.

Easterby-Smith, M. (1991). Management Research: An Introduction. London: Sage Publications.

Eastman, C. (1970). On the analysis of intuitive design processes. In G. T. Moore (Ed.), *Emerging methods in environmental design and planning. Proceedings of the First International Design Methods Conference* (pp. 21-37). Cambridge, MA: MIT Press.

Epstein, S. & Pacini, R. (1999). Some basic issues regarding dual process theories from the perspective of cognitive-experiential self-theory. In S. Chaiken & Y. Trope (Eds.), *Dual-process theories in social psychology*. New York: Guilford.

Feshbach, S. (1976). The Role of Fantasy in the Response to Television. *The Journal of Social Issues, 32* (4), 71-85.

Field, P. M, & Morse, J. (1985). Nursing Research: The application of qualitative approach. London: Chapman and Hall.

Fielden G. (1963). *Engineering Design*, 1963 (Report of Royal Commission, London).

Finke, R.A. (1989). Principles of Mental Imagery. Cambridge, MA: MIT Press.

Fish, J.M., & Scrivener, S. (1990). Amplifying the mind's eye: sketching and visual cognition. Leonardo.

Fodor, J. A. (1975). The Language of Thought. New York: Thomas Crowell.

Folkmann, M. N. (2010). Enabling creativity. Imagination in design processes. In *1st International Conference on Design Creativity ICDC.*

Forbus, K. D. (1984). Qualitative process theory. *Artificial Intelligence*, 24:85-168.

Forbus, K. M., & Gentner, D. (1997). Qualitative mental models: Simulations or memories? Proceedings of the *Eleventh International Workshop on Qualitative Reasoning*, Cortona, Italy, June 3-6, pp. 97-104.

Forbus, K., Gentner, D., Markman, A. & Ferguson, R.(1997). Analogy just looks like high-level perception: Why a domain-general approach to analogical mapping is right. *Journal of Experimental and Theoretical Artificial Intelligence (JETI)* 4, 185-211.

Fossey, E., Harvey, C., McDermott, F., & Davidson, L. (2002). Understanding and evaluating qualitative research. *Australian and New Zealand Journal of Psychiatry, 36*, 717-732.

Freeman, J. (2001). 2nd and 3rd International Workshops on Presence. *Presence: Teleoperators & Virtual Environments, 10*(3), 3-5.

Freeman, J., Avons, S., Meddis, R., Pearson, D., & IJsselsteijn, W. (2000). Using behavioural realism to estimate presence: A study of the utility of postural responses to motion stimuli. *Presence: Teleoperators and Virtual Environments. Presence*, 9, 149-164.

Frosch, D. L., Krueger, P., Hornik, R., Barg, F., & Cronholm, P. (2007). Creating demand for prescription drugs: a content analysis of television direct-to-consumer-advertising. *Ann Fam Med.* 2007;5(1):6-13.

Garau, M. (2003). *The impact of avatar fidelity on social interaction in virtual environments.* Unpublished doctoral dissertation, University College London.

Garau, M., Ritter-Widenfeld, H., Antley, A., Friedman, D.,Brogni, A., & Slater, M. (2008). Temporal and spatial variations in presence: A qualitative analysis. *Proceedings of the 7th Annual International Workshop on Presence PRESENCE 2004,* 232-239.

Gentner, D. (2002). Psychology of Mental models. In N. J. Smelser & P. B. Bates (Eds.), *International Encyclopedia of the Social and Behavioral Sciences* (pp. 9683-9687). Amsterdam: Elsevier Science.

Gentner, D. & Stevens, A. L. (Eds.). (1983). Mental models. Hillsdale, NJ: Lawrence Erlbaum Associates.

Gerrig, R. J. (1993). Experiencing Narrative Worlds. New Haven, CT: Yale University Press.

Gerrig, R. (1994). Narrative Thought? *Personality and Social Psychology Bulletin, 20* (6), 712–15.

Gerrig, R., & Pillow, B. (1998). Developmental Perspective on the Construction of Disbelief. In *Believed-In Imaginings: The Narrative Construction of Reality* (Eds., J. de Rivera, & T. Sarbin). pp. 101-119. Washington, D.C.: American Psychological Association.

Gibbs, G. R. (2002). Qualitative Data Analysis: Explorations with NVivo. Buckingham: Open University Press.

Gibbs, G (2006) 'Qualitative resources'. In: *ESRC Research Methods Festival, 17-20 July 2006*, St Catherine's College, Oxford.

Gibson, J. J. (1973). On the Concept of Formless Invariants. In Visual Perception, *Leonardo* 6, 43-45.

Gibson, J. J. (1979). The ecological approach to visual perception. Boston: Houghton Mifflin.

Glaser, (1978). Theoretical Sensitivity. Mill Valley, CA: The Sociology Press.

Glaser, B. G. (1992). Basics of Grounded Theory Analysis: Emergence vs Forcing, Mill Valley, CA: The Sociology Press.

Glaser, B. G. (1994). More Grounded Theory Methodology: a reader, Mill Valley, CA: Sociology Press.

Glaser, B. G. (1995). Grounded Theory: 1984-1994, Mill Valley, CA: Sociology Press.

Glaser, B. G. (1998). Doing grounded theory: issues and discussions. Mill Valley, CA: Sociology Press.

Glaser, B. G., & Strauss, A. L. (1967). The Discovery of Grounded Theory: Strategies for Qualitative Research. New York: Aldine Publishing Company.

Glenberg, A. M. (1997). What is memory for? *Behavioral and Brain Sciences, 20*, 1-55.

Goede, R., & Villers, C. D. (2003).'The applicability of Grounded Theory as Research Methodology in Studies on the use of Methodologies in IS Practices', ACM International Conference Proceeding Series; Volume 47, *Proceedings of the 2003 annual research conference of the South African institute of computer scientists and information technologists on Enablement through technology*, 2003, ACM, pp. 208-217.

Goel, V. (1995). *Sketches of thought.* Cambridge, MA: MIT Press.

Goffman, E. (1959). The Presentation of Self in Everyday Life. New York, NY: Doubleday.

Goffman, E. (1961). In D. Cressey, (Ed.), Asylums, Holt, Rienhart & Wintson, New York, NY.

Goffman, E. (1970). Strategic Interaction. Oxford: Basil Blackwell.

Goldschmidt, G. (1983). Doing Design, Making Architecture, *Journal of Architectural Education,* 37(1), 8-13.

Goldschmidt, G. (1991). The dialectics of sketching. *Creativity Research Journal* 4(2), 123-143.

Goldschmidt, G. (1994). On visual design thinking: the vis kids of architecture. *Design Studies, 15*(2), 158-174.

Goldschmidt, G. (2004). Design Representation: Private Process, Public Image. In G. Goldschmidt & W. L. Porter (Eds.), *Design Representation* (pp.203-217). London: Springer-Verlag.

Goodman, N. (1976). Languages of Art: An Approach to a Theory of Symbols, (2nd ed.). Indianapolis, IN: Hackett.

Goulding, C. (1999). *Grounded theory: Some reflections on paradigm, procedures and misconceptions.* Working paper series, WP006/99, Wolverhampton: University of Wolverhampton.

Goulding, C. (2002). Grounded theory: A practical guide for management, business and market researchers. Thousand Oaks CA.: Sage Publications.

Grbich, C. (1999). Qualitative research in health. Sydney: Allen & Unwin.

Green, M. C. (2005). Transportation into narrative worlds: Implications for the self. In A. Tesser, D. A. Stapel, & J. W. Wood (Eds.), *On building, defending and regulating the self: A psychological perspective,* New York: Psychology Press, pp.53-75.

Green, M. C., & Brock, T. C. (2000). The role of transportation in the persuasiveness of public narratives. *Journal of Personality and Social Psychology, 79*(5), 701-721.

Green, M. C., & Brock, T. C. (2002). In the mind's eye: Transportation-imagery model of narrative persuasion. In M. C. Green, J. J. Strange, & T. C. Brock (Eds.).*Narrative Impact: Social and Cognitive Foundations.*

Green, M. C., Brock, T. C., & Kaufman, G. F. (2004). Understanding media enjoyment: The role of transportation into narrative worlds. *Communication Theory,* 14 (4), 328-347.

Greeno, J. G. & Moore, J. L. (1993). Situativity and symbols: Response to Vera and Simon. *Cognitive Science, 17,* 49-59.

Grots, A. & Pratschke, M. (2009). 'Design Thinking—Kreativität als Methode', *Marketing Review* St. Gallen, 2-2009, pp. 18-23.

Guba, E., & Lincoln, Y. S. (1989). Fourth Generation Evaluation. Beverly Hills, CA: Sage.

Guba, E. G., & Lincoln, Y. S. (1994). Competing paradigms in qualitative research. In N. K. Denzin & Y. S. Lincoln (Eds.), *Handbook of qualitative research* (pp. 105-117). London: Sage.

Guindon, R., Krasner, H., & Curtis, B. (1987). Breakdowns and processes during the early activities of software design by professionals. In G. M. Olson, S. Sheppard, & E. Soloway (Eds.), *Empirical Studies of Programmers: Second workshop (ESP2).* Norwood, NJ: Ablex.

Gustafsson, E., Dellve, L., Edlund, M., & Hagberg, M. (2003). The use of information technology among young adults—Experience, attitudes and health beliefs. *Applied Ergonomics, 34,* 565-570.

Gysbers, A., Klimmt, C., Hartmann, T., Nosper, A., & Vorderer, P. (2004). Exploring the book problem: Text design, mental representations of space, and spatial presence in readers. In M. A. Raya & B. R. Solaz (Eds.), *Seventh Annual International Workshop: Presence 2004.* Universidad Politecnica de Valencia.

Harris, P. L., Brown, E., Marriott, C., Whittal, S., & Harmer, S. (1991). Monsters, ghosts and witches: testing the limits of the fantasy-reality distinction in young children. *British Journal of Developmental Psychology,* 9, 105-123.

Hayes, R., & Oppenheim, R. (1997). Constructivism: Reality is what you make it. In T. Sexton & B. Griffin (Eds.), *Constructivist thinking in counseling practice, research and training* (pp. 19-41). New York: Teachers College Press.

Heath, T. (1984). Method in Architecture, John Wiley & Sons Ltd., Chichester.

Heeter, C. (1992). Being There: The Subjective Experience of Presence. *Presence: Teleoperators and Virtual Environments. 1*(2).

Henwood, K. L. (1997). Qualitative inquiry: Perspectives, methods and psychology. In J. T. E. Richardson, (Ed.), *Handbook of Qualitative Research Methods for Psychology and the Social Sciences*, pp. 25-40. The British Psychological Society, Leicester, 1996.

Hodge, D. R., & Gillespie, D. (2003). Phrase completions: An alternative to likert scales. *Social Work Research, 27*, 45-55.

Hoffman, K. (1998). The Trance Workbook: Understanding & Using the Power of Altered States. New York: Sterling Publishing Co.

Holland, J. H., Holyoak, K. J., Nisbett, R. E., & Thagard, P. R. (1986). Induction: Processes of Inference, Learning and Discovery. Cambridge, MA: MIT Press.

Hoppe, M.J., Wells, E.A., Morrison , D.M., Gilmore, M.R. & Wilsdon, A. (1995). Using focus groups to discuss sensitive topics with children. *Evaluation Review 19* (1), 102-14.

Hu, J., Janse, M. & Kong, H. (2005). User experience evaluation of a distributed interactive movie. Available: http://www.idemployee.id.tue.nl/j.hu/publications/HCII2005_icecream.pdf

IJsselsteijn, W.A. (2002). Elements of a Multi-level Theory of Presence: Phenomenology, mental processing and neural correlates. *Proceedings of PRESENCE 2002*, pp. 245-259, Porto, Portugal, 9-11 October 2002.

IJsselsteijn, W. A., De Ridder, H., Freeman, J., & Avons, S. E. (2000). Presence: Concept, determinants and measurement. *Proceedings of the SPIE, 3959*, 520-529.

IJsselsteijn, W. A., de Ridder, H., Freeman, J. F., Avons, S. E., & Bouwhuis, D. (2001). Effects of stereoscopic presentation, image motion, and screen size on subjective and objective corroborative measures of presence. *Presence: Teleoperators and Virtual Environments.* Presence, 10, 298-311.

IJsselsteijn, W. A., & Riva, G. (2003). Being There: The experience of presence in mediated environments. In G. Riva, F. Davide, & W. A. IJsselsteijn, (Eds.), *Being There - Concepts, Effects and Measurements of User Presence in Synthetic Environments*, Amsterdam: IOS Press. pp. 3-16.

Jacobson, D. (2001). Presence Revisited: Imagination, Competence, and Activity in Text-Based Virtual Worlds. *CyberPsychology & Behaviour, 4*(6), 653-673.

Jacobson, D. (2002). On Theorizing Presence. *The Journal of Virtual Environment 6*(1).

Jennett, C., Cox, A. L. & Cairns, P. (2008). Being 'in the game'. The Philosophy of Computer Games 2008, Potsdam, Germany, 8-10 May, 2008.

Johnson-Laird, P. (1983). Mental models. Cambridge, UK: Cambridge University Press.

Jones, J. (1963). A Method of Systematic Design. In Jones, J. C. and Thornley, D. G. (eds), *Conference on Design Methods,* Oxford: Pergamon.

Jones, J. (1970). Design Methods. London: Wiley.

Jones, M. T. (2008). Found in translation: Structural and cognitive aspects of the adaptation of comic art to film. Dissertation. Temple University.

Jones, M. T. (2009). Found in Translation: Structural and Cognitive Aspects of the Adaptation of Comic Art to Film. VDM Publishing.

Jones, S. (2002). (Re)writing the word: Methodological strategies and issues in qualitative research. *Journal of College Student Development, 43*(4), 461-473.

Jones, S., & Hill, K. (2003). Understanding patterns of commitment: Student motivation for community service. *Journal of Higher Education, 74*(5), 516-539.

Kant, E. (1985). Understanding and automating algorithm design. *IEEE Transactions on Software Engineering, SE-11*, 1361-1374.

Kavakli, M., & Gero, J. S. (2001). Sketching as mental imagery processing. *Design Studies, 22*(4), 347-364.

Kebeck, G. (1997). Wahrnehmung. Theorien, Methoden und Forschungsergebnisse der Wahrnehmungspsychologie [Perception. Theories, methods and research findings of the psychology of perception]. Weinheim, München: Juventa.

Kember, D., Lai, T., Murphy, B., Siaw, I., Wong, J., & Tuen, K.S. (1990). *Naturalistic Evaluation of Distance Learning Courses.* Journal of Distance Education.

Kokotovich, V. (2000). Mental Synthesisand Creativity in Design: An Experimental

Examination, *Design Studies* 21(5): 437-449.

Kolko.J. (2010) Abductive Thinking and Sensemaking: the Drivers of Design Synthesis. *Design Issues* 26(1).

Kosslyn, S. M. (1978a). Measuring the Visual Angle of the Mind's Eye. *Cognitive Psychology* (10) 356-389.

Kosslyn, S. M. (1978b). Imagery and Internal Representation. In E. Rosch & B. B. Lloyd (Eds.), *Cognition and Categorization.* Hillsdale, NJ: Erlbaum.

Kosslyn, S. M. (1980). *Image and Mind.* Cambridge, MA: Harvard University Press.

Kosslyn S.M., Ganis G. & Thompson W.L. (2001). Neural foundations of imagery. *Nat. Rev. Neuroscience.* 2, 635–642.

Krippendorff, K. (2006). The semantic turn: A new foundation for design. Boca Raton: Taylor & Francis.

Krueger, R.A. (1988). Focus groups. A practical guide for applied research. Newbury Park, CA. Sage Publications.

Kuipers, B. (1994). *Qualitative Reasoning: Modeling and Simulation with Incomplete Knowledge.* Cambridge, Massachusetts: The MIT Press.

Kvale, S. (1996). Interviews: An introduction to qualitative research interviewing. UK: Sage Publications.

Lang, P. J. (1979). A bio-informational theory of emotional imagery. *Psychophysiology,* 16(6), 495-512.

Lankshear, A.J. (1993). The use of focus groups in a study of attitudes to student nurse assessment. *Journal of Advanced Nursing* 18, 1986-89.

Laseau, P. (1986). Graphic Problem Solving for Architects and Designers, (2nd ed.), New York: Van Nostrand Reinhold.

Lawson, B. (1979). Cognitive strategies in architectural design. *Ergonomics 22* (1), 59-68.

Lawson, B. (1980). How Designers Think. London: The Architectural Press.

Lawson, B. (2004). What Designers Know. Oxford: Architectural Press.

Lee, K. M. (2004). Presence, Explicated, *Communication Theory*, 14(1), 27-50.

Lessiter, J., Freeman, J., Keogh, E.,& Davidoff, J. (2001). A cross-media presence questionnaire: The ITC-Sense of Presence Inventory. *Presence: Teleoperators & Virtual Environments, 10* (5)3, 282-298.

Liddament, T. (2000). The myths of imagery. *Design Studies, 21*(6), 589-606.

Likert, R. (1932). A Technique for the Measurement of Attitudes, *Archives of Psychology*, No.140

Lombard, M. (2011). Why Telepresence Matters: A Person Answer to the "So What?" Question. Paper presented at the *International Society for Presence Research Annual Conference*, Edinburgh, Scotland.

Lombard, M., & Ditton, T. (1997). At the heart of it all: The concept of presence. *Journal of Computer-mediated communication, 3*(2).

Lombard, M. & Jones, M. T. (2007). Identifying the (Tele)Presence Literature. *Psychnology Journal, 5* (2), 197-206.

Lombard, M., & Jones, M. T. (2009). Defining presence. In F. Biocca, W. A. IJsselsteijn, J. Freeman, & M. Lombard (Eds.), Immersed in media I: Telepresence theory, measurement and technology. New York: Routledge.

Loomis, J. M. (1992). Presence and distal attribution: Phenomenology, determinants, and assessment. In *Proceedings of the SPIE, 1666,* 590-594.

Mackinder, M. & Marvin, H. (1982). Design: Decision Making in Architectural Practice. In *BRE Information Paper*, Ip 11/82, July 1982.

Madill, A., Jordan, A., & Shirley, C. (2000). Objectivity and reliability in qualitative analysis: Realist, contextualist and radical constructionist epistemologies. *British Psychological Society, 91*, 1-20.

Maher, M. L., Poon, J., & Boulanger, S. (1996). Formalising design exploration as co-evolution: A combined gene approach. In J. S. Gero, & F. Sudweeks (Eds.), *Advances in formal design methods for CAD* (pp. 1-28). London: Chapman & Hall.

Mantovani, G., & Riva, G. (1999). 'Real' presence: How Different Ontologies Generate Different Criteria for Presence, Telepresence and Virtual Presence. *Presence: Teleoperators and Virtual Environments, 8*(5), 538-548.

Markman, A. B., & Gentner, D. (2001). Thinking. *Annual Review of Psychology*, 52, 223-247.

Mason, J. (1996). Qualitative researching. London: Sage.

Maxwell, J. A. (1992). Understanding and validity in qualitative research. In A. M. Huberman, & M. B. Miles (Eds.), *The qualitative researcher's companion,* (pp. 37-64). Thousand Oaks, CA: Sage Publications (Reprinted from Harvard Educational Review. 1992, 62, 3; 279-300).

McCann, T., & Clark, E. (2003a). Grounded theory in nursing research: Part 2—Critique. *Nurse Researcher, 11*(2), 19-28.

McCann, T., & Clark, E. (2003b). Grounded theory in nursing research: Part 3—Application. *Nurse Researcher, 11*(2), 29-39.

McKellar, P. (1957). Imagination and Thinking. London: Cohen & West.

McKim, R.H. (1972). Experiences in visual thinking. Monterey, CA: Brookser Cole.

McKim, R. H. (1980). Experiences in Visual Thinking, Belmont: Wadsworth.

Mead, G. H. (1964). George Herbert Mead on social psychology. Edited by A. Strauss. Chicago: University Press.

Merrick, E. (1998). An exploration of quality in qualitative research: Are reliability and validity relevant? In M. Kopala, & L.A. Suzuki, (Eds.), *Using Qualitative Methods in Psychology*, pp. 25-36. Thousand Oaks, CA: Sage Publications.

Meyer J.H.F. and Land R. (Eds) (2006). Overcoming Barriers to Student Understanding: Threshold concepts and troublesome knowledge, London: Routledge.

Michalek, J. J., & Papalambros, P. Y. (2002). Interactive design optimization of architectural layouts. *Engineering Optimization, 34*(5), 485-501.

Mills, J., Bonner, A., & Francis, K. (2006). The development of constructivist grounded theory. International Journal of Qualitative Methods, 5(1).

Mills, J., Bonner, A., & Francis, K. (2006). Adopting a Constructivist Approach to Grounded Theory: Implications for Research Design. *International Journal of Nursing Practice, 12* (1), 8-13.

Minsky, M. (1980). Telepresence. *Omni*, June, 45–51. MIT Press Journals.

Mitchell, W. (1977). Computer-Aided Architectural Design. New York.

Mitchell, W. (1990). The Logic of Architecture. Design: Computation and Cognition. Cambridge, Mass.

Moore, K. (2003). *Overlooking the visual.* Routledge.

Morrison, M. (2002). What do we mean by educational research? In M. Coleman & A. Briggs (Eds.), *Research methods in educational leadership and management*, pp. 213-232. London: Sage.

Moseley, L. (1963). Rational Design Theory for Planning Building based on the Analysis and Solution of Circulation Problems. *Architects' Journal*, pp. 525-537.

Munhall, P. (2001). Ethical considerations in qualitative research. In P. Munhall (Ed.), *Nursing research: A qualitative perspective* (3rd ed., pp. 537-549). Sudbury, MA: Jones and Bartlett.

Nagai, Y. & Noguchi, I. (2003). An experimental study on the design thinking process started from difficult keywords: modelling the thinking process of creative design. *Journal of Engineering Design 14*(4), 429-437.

Nelson, M., & Poulin, K. (1997). Method of constructivist inquiry. In T. Sexton & B. Griffin (Eds.), *Constructivist thinking in counseling practice* (pp. 157-173). New York: Teachers College Press.

Nelson, H.G., & Stolterman, E. (2003). *The design way: Intentional change in an unpredictable world. Foundations and fundamentals of design competence.* Englewood Cliffs, NJ: Educational Technology Publications.

Newell, A., & Simon, H. A. (1972). *Human problem solving.* Englewood Cliffs, NJ: Prentice-Hall.

Neuman, W. L. (1991) Social Research Methods – Qualitative and Quantitative Approaches. Boston: Pearson Education

Norman, D. A. (1983). Some observations on mental models. In D. Gentner & A. Stevens (Eds.), *Mental models* (pp. 6-14). Hillsdale, N.J.: Lawrence Erlbaum Associates.

Norton, L. (1999). The philosophical bases of grounded theory and their implications for research practice. *Nurse Researcher, 7*(1), 31-43.

Nunez, D. (2007). A Capacity Limited, Cognitive Constructionist Model Of Virtual Presence. PhD Dissertation. University of Cape Town, South Africa.

Nunez, D. & Blake, E. (2001). Cognitive presence as a unified concept of virtual reality effectiveness. *Proceedings of AFRIGRAPH 2001*, 115-118.

Nunez, D., & Blake, E. H. (2003). Conceptual priming as a determinant of presence in virtual environments. Presented at the 2nd International Conference on Computer Graphics, *Virtual Reality, Visualisation and Interaction in Africa (AFRIGRAPH 2003)*, Cape Town, South Africa.

Nunez, D., & Blake, E. H. (2006). Content knowledge and thematic inertia predict virtual presence. Presented at PRESENCE 2005, the *9th International Workshop on Presence*, Cleveland, Ohio.

Oatley, K. (1994). A taxonomy of the emotions of literary response and a theory of identification in fictional narrative. *Poetics, 23*, 53-74.

Oatley, K. (1999). Why fiction may be twice as true as fact: Fiction as cognitive and emotional simulation. *Review of General Psychology, 3* (2), 101-117.

Ormerod, T. C. (2005). Planning and ill-defined problems. In R. Morris & G. Ward (Eds.), *The cognitive psychology of planning* (Vol. 1, pp. 53–70). London: Psychology Press.

O'Neill, S. & Benyon, D. (2003). A semiotic approach to investigating presence. COSIGN-2003, September 9-12, 2003.

O'Neill, S. J., McCall, R., Smyth, M., & Benyon, D. R. (2004). Probing the sense of place. *Seventh Annual International Workshop: Presence 2004.*

Owens, R. (1982). Methodological rigor in naturalistic inquiry: Some issues and answers. Educational Administration Quarterly, 18(2), pp. 1-21.

Oxman, R. E. (1994). Precedents in design: a computational model for the organization of precedent knowledge. *Design Studies* 12(2), 141-157.

Pace, S. (2003). '*Understanding the flow experiences of Web users*', PhD book, Australian National University, Canberra.

Pace, S. 2008. 'Immersion, flow and the experiences of game players', in E. Leigh, & R. Luechtefeld (Eds.), *SimTecT 2008 Conference Proceedings*, Simulation Industry Association of Australia, Lindfield, NSW, pp. 419-424.

Palmer, M. (1995). Interpersonal Communication and Virtual Reality: Mediating Interpersonal Relationships. In *Communication in the Age of Virtual Reality.* F. Biocca, & M. Levy (Eds.) (1995) Hillsdale, NJ: Lawrence Erlbaum Associates. pp. 277-299.

Palmgreen, P. (1971). A daydream model of communication. Association for Education in Journalism, Lexington, KY.

Pandit, N. R. (1996). The Creation of Theory: A Recent Application of the Grounded Theory Method. The Qualitative Report, 2(4), December, 1996.

Papantonopoulos, S. (2004). How system designers think: a study of design thinking in human factors engineering. *Ergonomics 47*(14), 1528-1548.

Paton, B. & Dorst, K. (2011). Briefing And Reframing: A Situated Practice. *Design Studies 32* (6), 573-587.

Patton, M. Q. (1990). Qualitative evaluation and research methods. (2nd Ed). Newbury Park, CA: Sage.

Peirce, C. (1998). On the Logic of Drawing History from Ancient Documents. In *The Essential Peirce: Selected Philosophical Writings, 1893–1913*, by Charles S. Peirce, edited by Peirce Edition Project (Bloomington: Indiana University Press, 1998).

Pendlebury, M. (1996). The role of imagination in perception. *South African Journal of Philosophy,*15(4), 133-139.

Phillips, M. (2000). The sadeian interface: Computers and catharsis. *Digital Creativity*,11(2), 75-87.

Piaget, J. (1971) *Mental Imagery in the Child: A Study of the Development of Imaginal Representation.*

Pidgeon, N. (1996). Grounded theory: Theoretical background. In J. T. E. Richardson, (Ed.), *Handbook of Qualitative Research Methods for Psychology and the Social Sciences*, pages 75–85. Leicester: The British Psychological Society.

Pidgeon, N., & Henwood, K. (1997). Using grounded theory in psychological research. In N. Hayes (Ed.), *Doing qualitative analysis in psychology* (pp. 245-273). Hove, UK: Psychology Press.

Pinchbeck, D., & Stevens, B. (2005). Schemata, narrative, and presence. In M. Slater (Ed.), Proceedings of *The 8th International Workshop on Presence.* University College London – Department of Computer Science, 227-230.

Plattner, H., Meinel, C., & Weinberg, U. (2009). Design Thinking, mv-verlag, Munich 2009.

Polaine, A. (2011). Design Research – A Failure of Imagination? Proceedings of the *1st International Symposium for Design Education Researchers, Cumulus Association/DRS SIG on Design Pedagogy.* Paris, France 18–19 May 2011

Poldma, T. (2009). Taking up space: Exploring the design process. *New* York: Fairchild.

Poldma, T. (2009). Experiential knowledge and rigour in research. Session Paper: developing knowledge through design research: understanding experiential knowledge as tacit. International Association of Societies of Design Research Conference 2009, *Proceedings of the IASDR20009 Conference*, October 18-22, Seoul, South Korea

Powell, R.A. & Single H.M. (1996). Focus groups. *International Journal of Quality in Health Care* 8 (5), 499-504.

Punch, K.F. (2005). Introduction to social research: Quantitative and qualitative approaches. (2nd ed.). London: Sage.

Pylyshyn, Z. W. (1973). What the mind's eye tells the mind's brain: a critique of mental imagery. *Psychol. Bull.* 80, 1–24.

Pylyshyn, Z. W. (2001). Is the imagery debate over? If so, what was it about? In E. Dupoux (Ed.), *Language, Brain, and Cognitive Development: Essays in Honor of Jacques Mehler.* Cambridge, MA: MIT Press.

Pylyshyn, Z. W. (2002a). Mental Imagery: In search of a theory. *Behavioral and Brain Sciences* (25) 157-182.

Pylyshyn, Z. W. (2002b). Stalking the elusive mental image screen (reply to commentaries). *Behavioral and Brain Sciences.* (25) 216-237.

Pylyshyn, Z. W. (2003a). Return of the Mental Image: Are There Pictures in the Brain? *Trends in Cognitive Sciences* (7) 113-118.

Pylyshyn, Z. W. (2003b). Seeing and Visualizing: It's Not What You Think. Cambridge, MA: MIT Press.

Pylyshyn, Z. W. (2003c). Explaining Mental Imagery: Now You See It, Now You Don't (Reply to Kosslyn et al.). *Trends in Cognitive Sciences* (7,3) 111-112.

Pylyshyn, Z. W. (2004). Mental Imagery. In R. L. Gregory (Ed.),*The Oxford Companion to the Mind* (2nd ed.), Oxford: Oxford University Press.

Raaijmakers, Q. A., Van Hoof, A., Hart, T., Verbogt, T. & Vollebergh, W. (2000). Adolescents Mid-Point Responses On Likert-Type Scale Items: Neutral Or Missing Values? *International Journal of Public Opinion Research, 12*, 208–217.

Radford, A. & Woodbury, R. (1991). Preliminary notes on computational design and the remaking of architectural education. In *The Technology of Design ANZAZcA Conference Proceedings*. Woodbury, G. (Ed) University of Adelaide, Adelaide (1991).

Rassin, E., Merkelbach, H., & Spaan, V. (2001). When dreams become a royal road to confusion: Realistic dreams, dissociation and fantasy proneness. *The Journal of Nervous and Mental Disease, 189* (7), 478-481.

Reed, K. (1991). Mastering fiction writing. Cincinnati: F&W Publications.

Regenbrecht, H., & Schubert, T. (2002). Real and illusory interactions enhance presence in virtual environments. *Presence: Teleoperators and Virtual Environments, 11*(4), 425 – 434.

Reitman, W. (1964). Heuristic decision procedures, open constraints, and the structure of ill-defined problems. In M. W. Shelley, & G. L. Bryan (Eds.), *Human judgments and optimality* (pp. 282-315). New York: Wiley.

Remenyi, D., Williams, B., Money, A., & Swartz, E. (1998). *Doing Research in Business and Management*, London: Sage Publications.

Rittel, H. (1972). On the Planning Crisis—Systems Analysis of the First and Second Generation, Bedrifsokonomen, 8, pp. 390-396.

Rittel, H. & Webber, M. M. (1973/1984). Planning problems are wicked problems. In N. Cross (Ed.), *Developments in design methodology* (pp. 135-144). Chichester, England: Wiley (Originally published as part of Dilemmas in a general theory of planning, *Policy Sciences*, 1973, *4*, 155-169).

Riva G. (2006). Virtual Reality. Wiley Encyclopedia of Biomedical Engineering, 2006 Volume 4, 117.

Riva, G. (2007). Virtual Reality and Telepresence. *Science 318*(5854), 1241-1242

Riva, G. (2008a). From Virtual to Real Body: Virtual Reality as Embodied Technology *Journal of Cybertherapy and Rehabilitation 1*(1), 7-22.

Riva G. (2008b) Enacting Interactivity: The role of presence. In Morganti, F., Carassa, A., Riva, G. (Eds.) *Enacting Intersubjectivity. A cognitive and social perspective on the study of interactions.* Amsterdam, IOS Press, 2008 (pp. 97-114).

Riva G. (2009). Is Presence a Technology Issue? Some insights from Cognitive Sciences. *Virtual Reality.* 2009, 13 (3): 159-169.

Riva, G. (2011). From the body to the tools: The roots of presence. Keynote address at the *International Society for Presence Research Annual Conference*, Edinburgh, Scotland.

Riva, G. & Waterworth, J. A. (2003). Presence and the Self: a cognitive neuroscience approach. *Presence-Connect, 3* (3).

Riva, G., Bacchetta, M., Cesa, G., Conti, S. & Molinari, E., (2003). Six month follow-up of in-patient experiential-cognitive therapy for binge eating disorders. *CyberPsychology & Behavior* 6, 251–258.

Riva G., Anguera M., Wiederhold B. & Mantovani F. (Eds.) (2006). From Commuication to Presence: Cognition, Emotions and Culture towards the Ultimate Communicative Experience. Amsterdam, IOS Press.

Roam, D. (2008). The Back of the Napkin: Solving Problems and Selling Ideas with Pictures. Portfolio, Penguin, NY.

Rodrigo, R. (2010). 'Design as Interpretation: Exploring Threshold Concepts in First Year Design education'. *Proceedings of ConnectED, 2nd International Conference on Design Education*, Sydney.

Roozenburg, N., & Cross, N. (1991). Models of the design process - integrating across the disciplines. In V. Hubka. Proceedings of ICED 91, Heurista, Zürich.

Rowe, P. G. (1987). *Design thinking*. Cambridge, MA: MIT Press.

Ryan, G. W., & Bernard, H. R. (2000). Data management and analysis method. In N. Denzin, & Y. Lincoln (Eds.), *Handbook of qualitative research* (pp. 769–802), Thousand Oaks: Sage Publications.

Ryan, M.-L. (1991). Possible worlds, artificial intelligence, and narrative theory. Bloomington, IN: Indiana University Press.

Ryan, M.-L. (2001). Narrative as Virtual Reality: Immersion and Interactivity in Literature and Electronic Media. Baltimore: The Johns Hopkins University Press.

Schiano, D.J., & White, S. (1998). First noble truth of cyberspace: people are people (even when they MOO). In: *Human factors in computing systems: CHI 98 Conference Proceedings*. (Eds., C.-M. Karat, A. Lund, J. Cortaz, and J. Karat). New York: Association for Computing Machinery, pp. 352–359.

Schlögl A., Neuper C. & Pfurtscheller G. (2002). Estimating the mutual information of an EEG-based Brain-Computer-Interface. *Biomedizinische Technik 47*(1-2), 3-8.

Schön, D. A. (1983). *The Reflective Practitioner*. New York: Basic Books.

Schön, D. (1984). Design: a process of enquiry, experimentation and research. *Design Studies 5* (3), 130-131.

Schubert, T. W. (2003). The sense of presence in virtual environments: A three-component scale measuring spatial presence, involvement, and realness. *Zeitschrift fur Medienpsychologie*, 15, 2, 69-71.

Schubert, T. (2009). A new conceptualization of spatial presence: Once again, with feeling. *Communication Theory 19*, 161-187.

Schubert, T., & Crusius, J. (2002). Five theses on the book problem: Presence in books, film and VR. Paper presented at Presence 2002 – *5th Annual International Workshop on Presence*. Porto, Portugal.

Schubert, T., Friedmann, F., & Regenbrecht, H. (2001). The experience of presence: Factor analytic insights. *Presence: Teleoperators and Virtual Environments, 10,* 266-281.

Schwandt, T.A. (1994). Constructivist, Interpretivist Approaches to Human Enquiry. In Denzin, N.K. and Lincoln, Y.S. (Eds) *Handbook of Qualitative Research*, Thousand Oaks, Sage.

Shaughnessy, J. J., & Zechmeister, E. B. (1997). Research methods in psychology. Singapore: McGraw Hill.

Sheridan, T. (1992). Musings on Telepresence and Virtual Presence. Telepresence. Presence: *Teleoperators and Virtual Environments, 1*(1),120-126.

Simon, H. A. (1969/1999). *The sciences of the artificial* (3rd, rev. ed. 1996) (3rd ed.). Cambridge, MA: MIT Press. (Original work published 1969).

Simon, H. A. (1971/1975). Style in design. In C. Eastman (Ed.), *Spatial synthesisin computer-aided building design* (pp. 287-309). London: Applied Science Publishers. First published in J. Archea, & C. Eastman (Eds.) (1971). *EDRA TWO, Proceedings of the 2nd Ann. Environmental Design Research*

Association Conference, 1-10, October 1970 (pp. 1-10). Stroudsbury, PA: Dowden, Hutchinson & Ross, Inc. (Version of Simon's text we refer to).

Simon, H. A. (1973/1984). The structure of ill-structured problems. *Artificial Intelligence, 4*, 181-201. Also in N. Cross (Ed.). (1984), *Developments in design methodology* (pp. 145-166). Chichester, England: Wiley.

Simon, H. A. (1987/1995). Problem forming, problem finding, and problem solving in design. In A. Collen, & W. W. Gasparski (Eds.), *Design and systems: General applications of methodology* (Vol. 3, pp. 245-257). New Brunswick, NJ: Transaction Publishers. (Text of a lecture delivered to the First International Congress on Planning and Design Theory, Boston, MA, 1987).

Simon, H. (1981). The Sciences of the Artificial. 2nd Ed. MIT Press, Cambridge, Mass.

Singh, A. (1999). The potential of mental imaging in the architectural design process. In *IDATER 99: International Conference on Design and Technology Educational Research and Curriculum Development.* Department of Design and Technology, Loughborough University.

Slater, M. (2003). A note on presence terminology. *Presence-Connect, 3* (3).

Slater, M. (2004). How colorful was your day? Why questionnaires cannot assess presence in virtual environments. *Presence: Teleoperators and Virtual Environments*, 13, 484-493.

Slater, M., Antley, A., Davison, A., Swapp, D., Guger, C., Barker, C., Pistrang, N., & Sanchez-Vives, M. V. (2006). A virtual reprise of the Stanley Milgram obedience experiments. PLoS ONE, 1, e39. doi: 0.1371/journal.pone.0000039.

Slater, M., Linakis, V., Usoh, M., & Kooper, R. (1996). Immersion, presence, and performance in virtual environments: An experiment with tri-dimensional chess. In M. Green (Ed), ACM Virtual Reality Software and Technology (pp. 163-172), 1-4 July 1996.

Slater, M., & Steed, A. (2000). A virtual presence counter. *Presence: Teleoperators and Virtual Environments*, 9, 413-434.

Slater, M., & Usoh, M. (1993). The influence of a virtual body on presence in virtual environments. *Presence: Teleoperators and Virtual Environments*, 3, 130–144.

Slater, M., Usoh, M., & Steed, A. (1994). Depth of presence in virtual environments. *Presence: Teleoperators and Virtual Environment 3*, 130-144.

Slater, M., Brogni, A., & Steed, A. (2003). Physiological Responses to Breaks in Presence: A Pilot Study. In *Proceedings of the 6th International Workshop on Presence.* Aalborg, Denmark, 6-8 October 2003.

Slater, M., & Wilbur S. (1997). A Framework for Immersive Virtual Environments (FIVE): Speculations on the Role of Presence in Virtual Environments. *Presence: Teleoperators and Virtual Environments. 6*(6).

Slater, M., Pérez-Marcos, D., Ehrsson, H., Sanchez-Vives, M., V, (2008). Towards a digital body: the virtual arm illusion. *Frontiers in Human Neuroscience.*

Smith, J. A. (1995). Semi-structured interviewing and qualitative analysis. In J. A. Smith, R. Harre, & L. Langenhove, (Eds.), *Rethinking Methods in Psychology*, pp. 9-26. London: Sage Publications.

Snodgrass, A., & Coyne, R. (1992). Models, metaphors and the hermeneutics of designing. *Design Issues*, 56-74.

Snodgrass, A., & Coyne, R. (1997). Is designing hermeneutical?. *Architectural Theory Review, 2*(1), 65-97.

Snodgrass, A. and R. Coyne. 2006. *Interpretation in Architecture: Design as a Way of Thinking.* London: Routledge.

Sommer, R. (1978). The mind's eye: imagery in everyday life. New York: Delta.

Stewart, S. (ed.) (2011). Interpreting Design Thinking. Special issue of *Design Studies 32* (6).

Stern, P. N. (1980). Grounded theory methodology: it uses and processes, *Image* 12 (1980), pp. 20-23.

Stern, P. N. (1994). Eroding Grounded Theory.. In J. Morse, (Ed.), *Critical Issues in Qualitative Research Methods*, Thousand Oaks: Sage.

Steuer J. (1992). Defining Virtual Reality: Dimensions Determining Telepresence. In *Communication in the Age of Virtual Reality*. F. Biocca, & M. Levy, (Eds.), (1995) Hillsdale, NJ: Lawrence Erlbaum Associates. pp. 33-56.

Stiles, W. B. (1993). Quality control in qualitative research. *Clinical Psychology Review*, 13:593-618, 1993.

Stratton, P. (1997). Attributional coding of interview data: Meeting the needs of long-haul passengers. In N. Hayes (Ed.), *Doing qualitative analysis in psychology* (pp. 115-141). Hove, UK: Psychology Press.

Strauss, A.L. (1987). Qualitative analysis for social scientists. Cambridge, MA: Cambridge University Press.

Strauss, A. L. (1991). Creating sociological awareness: Collective images and symbolic representation, Transaction Pubs: New Brunswick, NJ.

Strauss, A., & Corbin, J. (1990). *Basics of qualitative research: Grounded theory procedures and techniques.* Sage Publications.

Strauss, A., & Corbin, J. (1994). Grounded Theory methodology: An overview. In N. Denzin, & Y. Lincoln, (Eds.) *Handbook of Qualitative Research.* London: Sage Publications. pp. 1-18.

Strauss A., & Corbin J. (1997). Grounded Theory in Practice, Sage Publications, Inc.

Strauss, A., & Corbin, J. (1998). Basics of Qualitative Research: Grounded Theory Procedures and Techniques, (2nd ed.). Thousand Oaks CA: Sage. (First edition published in 1990).

Suwa, M., & Tversky, B. (1996). What architects see in their sketches: implications for design tools. In *Proceedings of CHI'96: Conference on Human Factors in Computing Systems*, pp. 191–192. New York: ACM Press.

Tafel, E. (1979). Years With Frank Lloyd Wright: Apprentice to Genius. Mineola, N.Y.: Dover Publications.

Tan, E. S. (1996). Emotion and the structure of narrative film. Film as an emotion machine. Mahwah: Lawrence Erlbaum.

Teal, R. (2010). Developing a (non-linear) practice of design thinking. *International Journal of Art & Design Education, 29*(3), 294-302.

Teeravarunyou, S. & Teixeira, C. (2002). *Visible Language 36* (2).

Thelen, E. & Smith, L. (1994). A dynamic systems approach to the development of cognition and action. Cambridge, MA: MIT Press

Thomsen, M. (2004). *Discovering mixed reality: Inventing design criteria for an action-based mixed reality.* Doctoral dissertation, University College London.

Thornley, D. G. (1963). Design Method in Architectural Education. In Jones, J. C. and D. G. Thornley, (eds.) (1963). *Conference on Design Methods.* Oxford, UK, Pergamon Press.

Tonkinwise, C. (2011). *Amplifying Creative Communities 2011 Northwest Brooklyn: Kinds and Product of Social Design, Part 1.Core 77*, 20 Dec 2011

Turner, P. (2009). The end of cognition? *Human Technology, 5* (1), 5-11.

Turner, P., & Turner, S. (2003). Two phenomenological studies of place. *People and Computers XVII-Proc. HCI Conference*, 21-35.

Turner, P. & Turner, S. (2006). Place, sense of place and presence. *Presence: Teleoperators and Virtual Environments, 15* (2), 204-217.

Turner, P. & Turner, S. (2011). The Book Problem Is All In The Mind. Paper presented at the *International Society for Presence Research Annual Conference*, Edinburgh, Scotland.

Turner, P., Turner, S., & Carroll, F., (2005). Spaces, Spatiality and Technology: The Tourist Gaze: Towards Contextualised Virtual Environments. *The Kluwer International Series on Computer Supported Cooperative Work Vol. 5.*

Turner, S., Turner, P., Carroll, F., O'Neill, S., Benyon, D., McCall, R., et al. (2003). Re-creating the Botanics: Towards a sense of place in virtual environments. *Environmental Science Conference.*

UK Design Council, 2009. http://www.designcouncil.org.uk/Documents/Documents/Publications/Research/Briefings/DesignCouncilBriefing04_DrivingRecoveryWithDesign.pdf

Ullman, D. G., Dietterich, T. G., & Staufer, L. A. (1988). A model of the mechanical design process based on empirical data. *AI EDAM, 2*, 33-52.

Visser, W. (1987a, 18-22 mai). *Abandon d'un plan hiérarchique dans une activité de conception.* Paper presented at COGNITIVA 87, Colloque scientifique MARI 87 Machines et Réseaux Intelligents (Tome 1), Paris, La Villette. English translation: W. Visser (1988). Gi*ving up a hierarchical plan in a design activity.* Rocquencourt, France: Institut National de Recherche en Informatique et en Automatique.

Visser, W. (1987b). Strategies in programming programmable controllers: A field study on a professional programmer. In G. M. Olson, S. Sheppard & E. Soloway (Eds.), *Empirical Studies of Programmers: Second workshop (ESP2)* (pp. 217-230). Norwood, NJ: Ablex.

Visser, W. (1991). Evocation and elaboration of solutions: Different types of problem-solving actions. An empirical study on the design of an aerospace artifact. In T. Kohonen, & F. Fogelman-Soulié (Eds.), *Cognitiva 90. At the crossroads of artificial intelligence, cognitive science and neuroscience. Proceedings of the third COGNITIVA symposium, Madrid, Spain, 20-23 November 1990* (pp. 689-696). Amsterdam: Elsevier.

Visser, W. (1995). Use of episodic knowledge and information in design problem solving. *Design Studies, 16*(2), 171-187. Also in N. Cross, H. Christiaans & K. Dorst (Eds.) (1996), *Analysing design activity* (Ch. 13, pp. 271-289). Chichester, England: Wiley.

Visser, W. (1996). Two functions of analogical reasoning in design: A cognitive-psychology approach. *Design Studies, 17*, 417-434.

Visser, W. (2006a, November 1-4). *Both generic design and different forms of designing.* Paper presented at Wonderground, the 2006 DRS (Design Research Society) International Conference, Lisbon, Portugal.

Visser, W. (2006b). *The cognitive artifacts of designing.* Mahwah, NJ: Lawrence Erlbaum Associates.

Visser, W. (2006c). Designing as construction of representations: A dynamic viewpoint in cognitive design research. *Human-Computer Interaction, Special Issue "Foundations of Design in HCI," 21*(1), 103-152.

Visser, W. (2009). Design: one, but in different forms. *Design Studies,* 30(3), 187-223.

Visser, W. (2010). Use of metaphoric gestures in an architectural design meeting: expressing the atmosphere of the building. *Proc. 4th Conf. Int. Society for Gesture Studies (ISGS),* p. 284.

Vorderer, P., Wirth, W., Saari, T., Gouveia, F. R., Biocca, F., Jancke, F., Bocking, S., Hartmann, T., Klimmt, C., Schramm, H., Laarni, J., Ravaja, N., Gouveia, L. B., Rebeiro, N., Sacau, A., Baumgartner, T., & Jancke, P. (2003). *Constructing Presence: Towards a two-level model of the formation of Spatial Presence.* Unpublished report to the European Community, IST Program, Project 'Presence: MEC' (IST-2001-37661). Hannover, Munich, Helsinki, Porto, Zurich.

Wade, J. (1977). Architecture, problems and purpose. New York: Wiley- Interscience.

Wallach, H., Safir, P. & Samana, R. (2010) Personality variables and presence. *Virtual Reality 14* (1), 3-13

Waterworth, E. L., & Waterworth, J. A. (2001). Focus, locus, and sensus: The three dimensions of virtual experience. *CyberPsychology & Behavior, 4*(2), 203-214.

Waterworth, J.A., & Waterworth, E.L. (2003a). Being and time: Judged presence and duration as a function of media form. *Presence: Teleoperators and Virtual Environments, 12*(5), 495-511.

Waterworth, J. A., & Waterworth, E. L. (2003b). The meaning of presence. *Presence-Connect,* 3, 3, posted 13-02-2003.

Waterworth, J. A. & Waterworth, E. (2003c). The core of presence: Presence as perceptual illusion. In *Presence-Connect, 3.*

Waterworth, E., Waterworth, J., Holmgren, J., Rimbark, T., & Lauria, R. (2001). The illusion of being present: Using the interactive tent to create immersive experiences. *Proceedings of Presence 2001, 4th International Workshop on Presence.* Philadelphia, May 21-23.

Weibel, D., Wissmath, B. & Mast, F. (2011). Cyberpsychology, Behavior, and Social Networking. October 2011, 14(10): 607-612.

Wertsch, J. V. (1998). Mediated action. In W. Bechtel & G. Graham (Eds.), *A Companion to Cognitive Science* (pp. 518-525). Malden, MA: Blackwell Publishers, Inc.

Willig, C. (2001). *Introducing Qualitative Research in Psychology: Adventures in Theory and Method.* Buckingham: Open University Press.

Wilson, M. (2002). Six views of embodied cognition. Available:http://philosophy.wisc.edu/shapiro/PHIL951articles/wilson.htm

Wirth, W., Hartmann, T., Bocking, S., Vorderer, P., Klimmt, C. & Schramm, H. (2007). A process model of the formation of spatial presence experiences. *Journal of Media Psychology*, 9, 493–525.

Witmer, B. G., & Singer, M. J. (1998). Measuring presence in virtual environments: A presence questionnaire. *Presence: Teleoperators and Virtual Environments, 7,* 225-240.

Witmer, B. G. & Singer, M. J. (2005). The Presence Questionnaire. *Presence: Teleoperators and Virtual Environments,* 14, 298–312.

Wu, W. (2011). What is conscious attention? *Philosophy and Phenomenological Research, 82* (1), 93-120.

Zeisel, J. (1981). *Inquiry by Design: tools for environment-behavior research.* Monterey, CA: Brooks and Cole.

Zeisel, J. (1984). *Inquiry by design.* Cambridge University Press

Zhao, S. (2003). Toward a taxonomy of copresence. Paper represented at the *4th Annual International Workshop on Presence.*

www.ingramcontent.com/pod-product-compliance
Ingram Content Group UK Ltd.
Pitfield, Milton Keynes, MK11 3LW, UK
UKHW041945190726
13854UKWH00004B/1802